Global Positioning System:

A Field Guide for the Social Sciences

John Spencer
University of North Carolina

Brian G. Frizzelle
University of North Carolina

Philip H. Page
University of North Carolina

John B. Vogler
East-West Center, Honolulu

 **Blackwell**
Publishing

350 Main Street, Malden, MA 02148-5020, USA
108 Cowley Road, Oxford OX4 1JF, UK
550 Swanston Street, Carlton, Victoria 3053, Australia

The right of John Spencer, Brian G. Frizzelle, Philip H. Page, and John B. Vogler to be identified as the Authors of this Work has been asserted in accordance with the UK Copyright, Designs, and Patents Act 1988.

First published 2003 by Blackwell Publishing Ltd

Library of Congress Cataloging-in-Publication Data

Global Positioning System : a field guide for the social sciences / John Spencer . . . [et al.].
    p. cm.
Includes bibliographical references and index.
    ISBN 1–4051–0184–9 (hardback : alk. paper) — ISBN 1–4051–0185–7 (pbk. : alk. paper)
    1. Global Positioning System.   I. Spencer, John.

G109.5 .G56 2003
910′.285—dc21   2002152561

ISBN 1-4051-0184-9 (hardback); ISBN 1-4051-0185-7 (paperback)

A catalogue record for this title is available from the British Library.

Set in 10.5 on 12.5 Times
by Ace Filmsetting Ltd, Frome, Somerset
Printed and bound in the United Kingdom
by MPG Books Ltd, Bodmin, Cornwall

For further information on
Blackwell Publishing, visit our website:
http://www.blackwellpublishing.com

Contents

Figures

Tables

Acknowledgments

This book represents the cumulative experience and knowledge of the authors as learned in the classroom and in the field. However, as with any learning process, there were many people along the way that served as mentors and guides. They helped us learn what should be done, what works and what doesn't, and each made a unique contribution that taught us a valuable lesson or provided guidance. There are two key people in particular the authors would like to collectively thank. Stephen J. Walsh from the Department of Geography at the University of North Carolina, Chapel Hill was instrumental in giving all of us the opportunity to develop our GPS and GIS skills either in the classroom, in the field or on the job.

Steve McGregor was the Associate Director for Spatial Analysis at the Carolina Population Center at the University of North Carolina, Chapel Hill. For the three of us who had the chance to work with him, he taught us many valuable lessons, including the importance of proper planning, proper GPS usage, and high quality in our work.

Through our work at the University of North Carolina, we have had opportunities to work with research fellows at the Carolina Population Center and faculty in the Department of Geography. Working with the following people gave us a chance to hone our skills and hopefully contribute to their research: Gustavo Angeles, Dick Bilsborrow, Ties Boerma, Barbara Entwisle, Wil Gesler, Ron Rindfuss, Amy Tsui, and Sharon Weir.

Individually, the authors would also like to thank the following people:

John Spencer: I would like to thank my parents for a lifetime of support and encouragement. Geoff Fuller gave extremely valuable assistance and counsel on the publishing world and was very instrumental in walking us through the process of finding a publisher. I don't know if this book would have happened at all without his advice. Brad Hutchinson, Jill Sunderlin, and Jill Sherman all reviewed drafts of chapters and gave many helpful comments. Melissa Conley-Spencer provided encouragement during the early stages of this work. Lastly, thanks to Kim Novak-Jones, Ken Jones, and Michael Jones for being there for me.

Brian Frizzelle: Thanks to my family for their never-ending love, support, and understanding, even if they don't fully comprehend what it is that I do. Yes, it does have something to do with geography, but not capitals of countries, and it also has to do with maps and satellites, but it's not "cartology" or "satellite surveying." Just as importantly, I would like to thank Katie Gaul for always being there for me, keeping me sane, and showing me what's truly important in life. Finally, to all of my friends, near and far, thank you for your support in this endeavor.

Phil Page: Thanks to Howard and Carol Page for their constant support, and Kelley Crews-Meyer for her words of encouragement and sage advice.

John Vogler: Thanks to my family, especially my mother and grandmother, for their constant love and support no matter where I go and what I do. I would also like to thank Annie "Annzies" Rohr for being my light at the end of the tunnel.

1
Introduction

Imagine a research project that is evaluating the availability of educational services in a developing country. As part of this project, villages will be visited and surveyed to determine the number of children in the village, how many of them can read or perform basic mathematics, which school they attend, as well as basic demographic information. As part of this effort, schools will also be visited to determine the number of teachers at the school, the subjects taught, how many students attend the school, and which village each student is from. All of the information listed above will provide a certain level of insight into the accessibility of education to the children in the villages. However, the investigators that are leading the project are interested in knowing how far each student has to travel to school. There is evidence suggesting that students who travel further than 1.5 km to school will be more likely to drop out, and the researchers would like to identify which villages are the farthest away from the schools.

The researcher's first idea is to ask the parents how far the children travel to school, however there is a significant body of literature that suggests people often do not accurately estimate distances they travel. The next thought is to use printed maps to determine where the villages are located, however upon investigation it is discovered that the existing maps have not been updated in over 15 years and many of the villages in the study area have developed since then. Another problem is that for those villages that are on the map, the schools aren't displayed, much less individual dwellings.

One day while pondering this problem, a colleague mentions to the investigator how global positioning system (GPS) was used in a project looking at health care accessibility. According to the colleague, GPS was cheap, easy to use, and provided very accurate results. GPS seems perfect for the education project, so the investigator decides to use it to obtain coordinates for the location of every household and school. These coordinates can then be used to calculate very precisely the distances children travel. Several GPS receivers are bought and passed out to members of the field team. They're told to read the manual and collect a coordinate whenever they visit a house or a school to administer the survey, and write the coordinate down on the survey.

Once the fieldwork is over, the project leader collects the completed surveys and goes to collect the GPS receivers. He discovers that one of them stopped working on the second day, another was lost, but the other receivers were used by the field team and coordinates are stored in the memory. As part of the data entry process of the survey, the coordinates are keyed in to a database. When the database is passed on to the Geographic Information System (GIS) Lab at his institution, a problem is discovered. Not every household or school has coordinates. Of those that do have coordinates, some are in the ocean; others are in Antarctica, while still others are in a district other than what is reported in the survey. Additionally, some coordinates are expressed in latitude and longitude, while others are in a national coordinate system. Given all of these problems, the researcher is told, it may not be possible to determine where the points should really be. In short, it is doubtful whether the points are useable at all.

The above scenario, while fictional, is loosely based on the experiences of an actual research project. It serves to illustrate the risks as well as the potential benefits of GPS. This book is written to help users maximize those benefits while avoiding the pitfalls. GPS is a deceptively simple technology: the receivers can be as small as a cell phone, cost less than US$100.00, and can provide a coordinate that is accurate to within 10 m with the press of a few buttons. However, like any tool used in the field, there must be planning and wise precautions to avoid wasting money and effort.

One of the purposes of this book is to provide detailed guidance to social science researchers who wish to incorporate GPS into their research, regardless of their discipline or past experience with geographic technologies such as GIS or GPS. While there is not a one-size-fits-all approach to using GPS in social science research, there are some key concepts and methods that are common to any project. Using GPS in a project requires more than just getting a GPS receiver, turning it on and writing down the number displayed on the screen.

Before going into the field, decisions must be made about the level of accuracy required to adequately answer the project's research question. Is 1 m accuracy needed or is 100 m accuracy sufficient? The answer will influence the types of receivers purchased, the data collection methodology as well as the GPS error correction methods employed. When in the field, how will the coordinates be collected, by whom, and what in-field verification procedures will exist? Once the fieldwork is complete how will the data be displayed or analyzed? These are just some of the questions that need to be addressed before fieldwork begins. There is a great value to asking these questions before going into the field and this lesson is an important one to take away from this book. An additional goal of this book is to provide the information necessary to allow the researcher to effectively plan and execute a GPS component in their research.

There has been a revolution in GPS over the last several years, as costs of receivers have decreased and accuracy has improved. GPS has become a critical tool in a variety of settings: aviation, shipping, emergency vehicle routing, even television and film production. The same characteristics that make it appealing to these users – reliability, cost, and ease of use, make it well suited for social science research. But perhaps the biggest boon to the use of GPS was the lifting of Selective Availability (S/A) by President Bill Clinton in May 2000. Selective Availability was the intentional error introduced in civilian receivers. When S/A was present, military users could get accuracy of 5 m or less, while civilian users typically could expect at least 100 m errors. At that time the only way to improve the accuracy of the coordinate was through the use of error correction techniques that were often costly and added a layer of complexity to using GPS. Citing the value of GPS to the civilian community, President Clinton's executive order lifted the restrictions on the signal, providing civilian users greater accuracy than previously possible. Given the economic impact reinstating S/A would have, it is unlikely that it will ever be turned back on across the entire system. In the future, any degradation of the signal will be isolated to parts of the world where there are US military actions or activities. It should be noted, however, that it is not clear at this stage whether the US government will inform users in these areas whether S/A is active.

Before beginning it is worthwhile to discuss the structure of this field guide. Conceptually, the book is divided into two sections. The first section, Understanding GPS, is written to provide the reader an understanding of the technology and the fundamental knowledge necessary to use it. The second section, Utilizing GPS, provides practical information on how to use the technology. The chapters in this section contain an overview of the planning and preparations necessary before going into the field, the work that should be done in the field, and how to make best use of the collected data.

After this introduction, Chapter 2 discusses the role that GPS can play in social science research and provides an overview of linking the data to a Geographic Information System (GIS). Examples are given of research projects that might use GPS. The chapter concludes by summarizing alternative ways besides GPS for locating phenomena and people.

Chapter 3 provides an overview of the technology itself. The history of the system as well as its current organization and administrative structure are discussed. While the actual concepts behind the system rely on the basic *distance* = *rate* × *time* formula, the technology at work adds a layer of complexity. This chapter discusses both the technology as well as the methods used by the receiver to calculate coordinates.

Basic concepts of geodesy are presented in Chapter 4. In order to most effectively use GPS it is necessary to have a basic understanding of such terms as coordinate system and datum. While it is possible to devote a whole book to these subjects, only those elements most relevant to the use of GPS are discussed. Coordinates returned by the GPS provide an absolute location on the earth's surface that is a part of a specific larger coordinate system. There are many coordinate systems available to the GPS user and determining the appropriate one will facilitate analysis. Datums, which are unfamiliar to many GPS users, are a critical factor to understand in order to use GPS effectively. While it can be difficult to succinctly define what datums are, they refer to the linking of the idealized shape of the earth to the actual shape of the earth. This level of discussion about the true shape of the earth compared to the idealized shape may seem like a rather esoteric topic and not one directly relevant to GPS, but not being aware of the datum being used can result in significant error.

The accuracy of the coordinates provided by GPS is discussed in Chapter 5. There are several potential sources of error present when using GPS. Some of these are systematic, while others are environmental. Regardless of the source, they can contribute to inaccurate coordinate calculations, and techniques to correct or minimize these errors are discussed. This chapter also discusses two different types of GPS receivers, mapping grade and recreational. Receiver features are detailed and common GPS related hardware is also presented. The chapter concludes by discussing the steps necessary for collecting a coordinate with a GPS receiver.

Part II, Utilizing GPS, focuses on the fieldwork stage itself. Because a GPS project can be a rather complex undertaking, there are many steps involved in the planning and execution of data collection, and the processing of the collected data. This part of the book breaks these steps down into manageable pieces. It is important to note however, that like any complicated undertaking it can be difficult to present the steps involved in a strictly linear fashion. Often going from step one to step two requires a side discussion about an issue that may have been presented in a previous chapter. When possible, the authors have attempted to simply refer the reader back to the appropriate chapter, however in some instances clarity and the importance of the issue dictated that it be presented again, albeit in a limited way.

Chapter 6 provides a comprehensive overview of the entire process of developing a GPS project. While the specific steps and requirements will vary from project to project, there are key steps and tasks that are consistent in all situations. It is important to define the goals of the project and understand the research objectives being addressed. It is also critical to prepare a detailed project plan that defines the characteristics of the data and the protocols for data collection in the field. Decisions on personnel, equipment, field instruments, and training are also vital at this stage. Once these issues are developed, it is important to resolve data quality and logistical issues, such as data verification in the field, transportation, and equipment troubleshooting. After the data has been collected, it is important to then have a plan for data processing and analysis. These topics are outlined in this chapter and presented in a cohesive manner, giving the reader an overarching view of a GPS project. The specifics of each step are presented in detail in subsequent chapters.

Planning and preparation are crucial to ensuring a successful GPS project. Chapter 7 discusses in detail the necessary tasks. In addition to reviewing the research question and project goals, it is important to determine the accuracy needs of the project as well as the coordinate system and datum that will be used in the study.

Depending on the goals and objectives of the GPS project, the planning may seem as complex as the planning done for a major military operation, and Chapters 8, 9, and 10 are accordingly detailed. Data and methods are addressed in Chapter 8. Section one discusses the characteristics of the spatial data, that is, what real landscape features and phenomena (natural and cultural) are being located, described, and represented in the spatial database. Section two addresses the development of a data collection methodology that satisfies a project's goals and accuracy requirements and results in data that is compatible with a project's existing spatial information.

In Chapter 9, field resources are addressed, including equipment, personnel, and field instruments. These can be the most costly aspects of the data collection effort. Equipment recommendations and descriptions are presented as well as pertinent personnel roles and their descriptions. Various field instruments, such as data collection forms, field maps, and reference lists that might be required to complete the fieldwork are discussed with some examples of data collection forms provided.

One goal of any research project is the compilation of high-quality data in the most efficient manner possible. To that end, data quality measures and fieldwork logistics are presented together in Chapter 10. The first part of the chapter discusses the refinement of data collection protocols and field instruments as well as ways to test the repeatability of procedures and enhance data quality. The second part of the chapter relates a set of logistical issues that, when addressed, should smooth the transition to the field, give the field season cohesion and get the researcher thinking beyond the fieldwork. The chapter ends with a summary of fieldwork planning and preparations.

When the fieldwork actually gets underway, the first step is to train those who will be using the GPS receivers. Chapter 11 discusses the training necessary for a successful project and provides an overview of the material that should be presented to the field teams. After training is completed, the data collection can begin. If the project is large enough in scope, territories, and work schedules of the field teams will need to be coordinated. The chapter also discusses GPS practicalities such as initializing, configuring, and recording data with the GPS receiver.

After fieldwork is complete, it is time to process and clean the data, and these steps are presented in Chapter 12. In addition to basic data cleaning and verification steps such as checking for out-of-range values and identifying missing data, the steps necessary for differential correction, a powerful error correction method are also discussed.

Geographic information systems are the most commonly used tool for reaping the full benefit from GPS data. Chapter 13 discusses ways to integrate GPS data into GIS-based analyses.

It is important to emphasize that every GPS project is different and some projects may have a large collection team dispersed widely throughout a country, while other projects may just consist of one person with one receiver collecting a handful of coordinates. What is important is not the number of people or the number of points being collected; instead developing an adequate, well-thought out field plan is a key part for any GPS project.

Lastly, it is important to note that specific manufacturers are mentioned by name in the book. This is not an endorsement of particular brands of receivers or software, but it is a necessity at times because some items have specific traits or characteristics that are unique and are mentioned in the text. A list of GPS manufacturers is provided in Appendix A.

Part I
Understanding GPS

2
Why use GPS?

What GPS Provides

Social scientists work at a variety of scales, and none of those is more complex than the scale of the individual human or household. It is at this scale that "observations and measurements of features, patterns and events are made by individuals" (Sexton et al., 1998, p. 181), and where human behavior is best documented. Many social science research projects involve detailed data collection and analysis at this scale, but when the spatial dimension is included in the study, the complexity increases. Consider a project in which the researcher is looking at socioeconomic data at the census block level. These data are aggregated to an area, thus eliminating the need to worry about precise locations of the subjects. But when it becomes necessary to locate individual people, residences, facilities, or any other single place, there is no easier way of doing so than using the global positioning system (GPS).

As we will see in this chapter, collection of locational data is important because the spatial arrangement, dispersion, and interaction of phenomena can be a valuable component of research. Thus, the global positioning system is a valuable resource for social scientists. It allows researchers to record locations of people, phenomena, buildings, and other objects of interest with little effort and minimal cost. It is a much more time- and cost-efficient means of recording positional data compared to alternative methods, such as traditional land surveying. It is also more accurate than approximating the coordinates of places or objects using hardcopy maps. Through the use of GPS technology, any research project can easily add a geographic component, opening the door to a wide array of spatial analytical methods.

A GPS receiver alone is not a panacea for answering geographic research questions. It is simply another piece of equipment for collecting data in the field. At its most basic, it collects a coordinate pair (e.g. latitude and longitude) for the location in which the user is standing at that particular time. This location is realized as a point when placed on a map. However, merely having the coordinates not only allows the user to accurately place that location on a map but to also link attributes about that location on a map as well. The accuracy provided by GPS is the true benefit of this technology. More complex applications of GPS allow the user to collect accurate locational data along a line or around the perimeter of an area, such as a lake or agricultural field. These data, when placed on a map, appear as lines or polygons (Kevany, 1994). Moreover, elevation measurements can be made using GPS, thus adding a third dimension (z) to point, line, and area data.

There are many potential applications for GPS within the social sciences. This technology can help the health care analyst locate health clinics, which is crucial for assessing the accessibility and delivery of health care to its nearby population. The sociologist can use GPS to locate recreational facilities and neighborhoods in a city in order to better understand how their proximity to neighborhoods affects adolescent fitness. The anthropologist can use GPS to locate archaeological sites, cultural landmarks, or other man-made features that are of interest. More specific examples of GPS use in various fields of research will be discussed later in the chapter.

Using GPS in social science research has many advantages. GPS is a relatively easy technology to use and is also very cost-effective. The accuracy obtainable using a low-cost, recreational grade receiver can be sufficient for many purposes. GPS receivers can be purchased for as little as US$100, and most are very user-friendly (see Chapter 5 for a discussion of different receiver types). Simplicity and affordability make it very feasible to incorporate GPS data collection into surveys and other fieldwork. However, the ease of use of this technology can be deceptive. Without adequate planning and training, the data collection efforts could result in data containing positional coordinates that require considerable cleaning or, in the worst case, are simply unusable. There are many considerations to take into account when collecting GPS data, and these will be discussed in more detail in later chapters.

Since proper data collection procedures are crucial for accurate readings, it is vitally important that users understand how GPS functions. In Chapter 3, you will learn how a GPS receiver uses the radio signals emitted from the satellites to calculate a set of coordinates. The method by which a position is located in space is conceptually quite simple. However, the manner in which the global positioning system functions is complex, and while the reliability of the system is high, it is not infallible. Although the receiver does most of the work on the user's end, it is still important for the user to be familiar with both the concept and the complexity. Failure to understand the workings of the technology can result in a "black box syndrome," where the user is unaware of the quality of the data output by the receiver. More specifically, this syndrome is characterized by the user expecting the receiver to always function correctly, thereby failing to recognize when there are problems with the system and taking the necessary measures, if within the user's control, to ensure that accurate data are collected. Part II of this book will take you through all of the steps for incorporating GPS into a research project, from determining goals and accuracy requirements and developing protocols, to collecting the data, to processing the data after fieldwork is complete, and integrating the data within a geographic information system (GIS).

Role of the Spatial Perspective in Social Science

Within the social sciences, the spatial perspective can explain a significant component of human activity, behavior, and decision-making. But in order to utilize the spatial perspective, one must have spatial locations for the phenomena of interest–in this case, people and their infrastructure, or activity spaces. It is not uncommon for social scientists to conduct research without considering where people and their activities are located in space. And while location is not always a necessity, it can have a considerable effect on any data that are being analyzed. In fact, the First Law of Geography (Tobler, 1970) states that all objects are related to one another, but those near one another are more similar than those farther apart. This "law" is more of an observation of the relationships that typically exist between phenomena based on their proximity to one another, and it is not a hard and fast rule. However, it does illustrate that the addition of the spatial perspective to any research will open the door to those associations that are based on location.

The incorporation of the spatial perspective has grown in the social sciences over the past several decades. With increasing computing power and decreasing cost of computers, as well as the continuing improvements in GIS software and advancement of spatial statistics, more traditional research methodologies have adopted the spatial perspective to better study the effects of location on different phenomena. However, it is only since the early 1990s that the use of GPS has become an important tool for capturing the locations of those phenomena (for an example, see Lang and Speed, 1990). In a similar fashion to computers and GIS software, GPS technology has continuously become more accessible and affordable, allowing researchers to more easily incorporate it into their work. A wide variety of research fields have made use of GPS technology to gather locational coordinates of the people, places, or events in which they are interested. These coordinates are most often captured and saved in a digital form, and then imported into a GIS to be used with other spatial data for analysis. Within the GIS, a researcher can attach attribute data about a location to a point, line, or area on a digital map, which allows for a variety of spatial analyses.

Uses of the global positioning system

There are many different uses for GPS technology, but here we are only interested in how it can be used to benefit research in the social sciences. We have broken down GPS use into three different categories: (1) mapping, (2) spatial analysis, and (3) ground truthing. Each of these will be briefly discussed in the next three sections in order to provide some insight into these applications. A subsequent section will include specific examples of GPS applications in several different fields of social science research.

GPS for mapping

The most basic use of GPS is for mapping. Many people are simply interested in locating phenomena that they want to place on a map, and GPS is a wonderful tool for doing so. It can be used to update reference maps (Kevany, 1994), or to map objects whose locations were previously unknown or inaccurate (Rybaczuk, 2001). Other times, these maps can show the distribution of some attribute of the phenomena at which the coordinates were collected, and the map then becomes an instrument for visual analysis of data. This is highly useful as a preliminary analysis tool, since the researcher can often see patterns on a map and begin to understand how the spatial distribution affects certain attributes (Priskin, 2001).

It is important to realize that GPS is most often used to collect coordinates of individual locations, which are realized as points in a geographic information system. However, some GPS receivers can also be used to collect and generate line and area (or polygon) data. This can be done in a variety of ways. One method is to collect discrete points along the length of a line (i.e. positions are recorded at each break or curve in the line) or at each corner of an area. The points can then be entered in a GIS and manually connected by straight-line segments to form the line or polygon feature. A second method, if the receiver has this capability, is to turn the receiver on and continuously record positions while traversing the length of the line or the perimeter of the area. This data can be entered into the GIS and automatically displayed as a line or polygon. This second method is simpler and will more accurately reproduce the overall shape of the line. However, when viewed at a very large scale, what should be a perfectly straight line in reality may appear to zigzag due to small errors in each coordinate pair that is calculated along that line (see Chapter 5 for a discussion of error in GPS). Another method allows the user to create area data from a single point. A point is collected on the ground within an area of interest (e.g. forest stand, park, land parcel). The point is then overlaid on top of contextual data, such as an aerial photograph or satellite image, within a GIS. This allows the user to positively identify and outline the area. A related use of GPS is for navigational purposes. For example, if a researcher identifies areas of interest prior to going to the field and can generate semiaccurate coordinates (Frizzelle and McGregor, 1999; Moran et al., 2000), or if a researcher wants to return to locations previously visited for which coordinates exist, a GPS receiver can be used to help the field team navigate to those locations.

GPS for spatial analysis

The next logical step for mapped point data locations is to perform spatial analyses on the data. The simplest form of spatial analysis is a point-pattern analysis, in which the researcher uses computer algorithms to determine if the point locations are clustered, evenly distributed, or randomly distributed. This tends to be a preliminary analysis, and does not require the researcher to have any additional data for the points other than their coordinates.

A more advanced use of point locations is the linking of aspatial, or nonspatial, attribute data for analysis. In fact, within the social sciences this is becoming one of the most common applications of GPS data in research projects. More and more studies are choosing to analyze their data within the context of its surrounding environment. GPS is the fastest and most accurate method of realizing this. Social scientists are utilizing GPS to link sociodemographic survey data to household point locations, health care data to clinic or hospital locations, and community characteristics to points that represent the center of the community. Environmental scientists are using GPS to locate field plots and analyze the characteristics of the vegetation (Halme and Tomppo, 2001), find animal roosts and examine the spatial determinants of roost locations based on the surrounding landscape (Lumsden et al., 2002), and identify locations of fishing debris in the ocean (Donohue et al., 2001). And the growing interdisciplinary field of population-environment research is utilizing GPS to more accurately link people to their land (Fox et al., 1996; Frizzelle and McGregor, 1999; Kammerbauer and Ardon, 1999; McCracken et al., 1999; Moran et al., 2000).

When aspatial attribute data are attached to point locations, lines, and/or polygons, a wide variety of GIS-based spatial analyses become available. These analyses allow the user to analyze any of the attached attribute variables within geographic space. Methods of linking these data and performing analyses within a GIS will be discussed in Chapter 13.

GPS for ground truthing

The third category of GPS use is ground truthing. Ground truthing is the collection of locations and corresponding information about features on the ground that will be used to create, correct, interpret, assess accuracy or somehow modify existing geospatial data. Two common uses of ground truth data are for georeferencing aerial photographs or satellite images and classifying satellite images for deriving land use and land cover (LULC). Ground truthing differs from the previous two uses of GPS in that the data collected are primarily used to augment processes that are performed on other geospatial data rather than being used for mapping or direct input into spatial analyses.

GPS data collected for georeferencing purposes is called *geodetic control* (Kammerbauer and Ardon, 1999; Moran et al., 2000). The collection of geodetic control points generates a control network that can be used as a basis for an accurate referencing system for a spatial database (Kevany, 1994). Geodetic control is commonly collected for use with digital aerial photos and satellite images. The main goal is to collect GPS control points at locations that are static and easily recognizable in the image, such as road intersections. Then the coordinates can be applied to the pixels of those static features in the image, a transformation algorithm is calculated and applied to the image, and the image is transformed so that every pixel in the new image has accurate coordinates.

Ground truthing also includes using GPS for the collection of land use and land cover data points for the interpretation and classification of multispectral satellite imagery (Moran, 2000). It is beyond the scope of this book to describe the different types of classifications, or the steps involved in a classification. However, it is necessary to mention that point locations with in situ data, associated attributes describing the type of LULC at those locations, can be related to particular pixels (pixel brightness values in various regions of the electromagnetic spectrum) in a satellite image. Once the researcher has a sufficient number of these points, the in situ LULC information can be used to create a model that will analyze the multispectral data and assign a class to each pixel.

Fields of research that incorporate GPS

The three general uses described above are utilized quite extensively in a variety of research fields. Even though the focus of this book is on the social sciences, this section will include some brief discussions of research in the environmental sciences to help illustrate the usefulness of GPS. This small foray into the environmental sciences is also relevant in that GPS is being widely used in interdisciplinary research between the social and physical sciences.

Investigators in sociology, demography, human geography, economics, anthropology, public health, urban and regional planning, tourism, and several other fields are using GPS technology within the social sciences. Similarly, environmental science researchers in ecology, geology, physical geography, marine and atmospheric sciences, animal behavior, and more are adopting it for their work. Outside of academia, it is in widespread use by government divisions at all levels, such as transportation and planning departments and the US Department of the Interior, and nongovernmental organizations and utility companies for improving their data infrastructure and modernizing their normal procedures.

Although they have begun to adopt this technology later than their colleagues in the environmental sciences, social scientists are some of the most avid academic users of GPS equipment. The integration of GPS into what traditionally had been nonspatial research methods has opened the door to a wide variety of new exploratory analytical techniques that many times allow for better understanding of the underlying relationships between people and their socioeconomic and demographic characteristics. Sociologists are using GPS to locate where people live, and using those locations to improve their knowledge of social phenomena such as social networks. The field of public health and other similar fields that study accessibility to health care and the effectiveness of health care programs are widely accepting GPS to allow them to better understand where people live in relationship to their health care providers. The field of transportation studies has eagerly adopted GPS technology for many applications, from locating features of road networks and improving general reference maps, to inventorying and repairing road damage, to examining causes of accidents. And GPS is being used quite extensively in social projects and programs by government and nongovernment organizations.

BOX 2.1

Examples of Social Science Research Utilizing GPS Technology

On the socioeconomic side, Faust et al. (1999) use GPS to locate villages in Nang Rong, Thailand, for pattern analysis. Village locations were linked to social survey data in a GIS, leading to the analysis of patterns of exchange of agricultural equipment and labor between villages, as well as movement of the populace to temples and schools. The Bureau of Indian Affairs has used GPS in Montana to locate all components of the irrigation network on the Blackfeet Reservation (Seagle and Bagwell, 2001). These point and line data were then incorporated into a GIS and used to create an efficient system for maintaining and upgrading the irrigation system for the Blackfeet people.

BOX 2.2

Examples of Health Care Research Utilizing GPS Technology

Health care researchers, such as medical geographers, use GPS extensively to locate people, health care facilities, and areas of disease outbreaks. After the implementation and expansion of a community-wide tuberculosis treatment program in Hlabisa, South Africa, Wilkinson and Tanser (1999) used GPS to document the increase in accessibility to treatment. To do so, they collected GPS points at households, health facilities, homes of community health workers and potential volunteer treatment supervisors, and performed the analyses within a GIS. Perry and Gesler (2000) used a GIS to assess physical access to primary health care in a remote, mountainous region of Bolivia. Community locations were recorded with a GPS unit, and then utilized in the analysis to determine which communities were within a 1-hour walk of a health facility. Ali et al. (2001) collected GPS locations of hospitals, treatment centers, village boundaries, religious institutions, schools, and household clusters in order to better understand the effects of health care provision on acute lower respiratory infection mortality in young Bangladeshi children.

BOX 2.3

Examples of GPS Utilization in Transportation Studies

Outside of the social, physical, and environmental sciences, the largest area of study that uses GPS is the transportation sector. GPS is used to assist those in road maintenance to the study of accident patterns to the identification of areas of high traffic congestion. Lee and Mannering (2002) used GPS to identify locations of "run-off-roadway accidents." They linked attribute information describing the road conditions, and analyzed the relationships between different road conditions and number of accidents in an integrated GIS.

Many of the applications for GPS in the physical and environmental sciences are similar to those in the social sciences, with one main difference: they rarely involve locations of humans. GPS technology is used extensively in geology for monitoring geologic shifts. It is also used in ecological studies for locating study sites, such as vegetation samples, and for tracking the movements of animals.

A cross-disciplinary field that is steadily implementing the use of GPS in research is the area of population–environment research. Population–environment research attempts to examine the interactions between humans and their environment, to better understand how human activities affect the landscape around them, and how that surrounding landscape in turn constrains or directs their actions. GPS allows the population–environment researcher to place humans accurately on the landscape. Whether the study is examining how humans utilize urban spaces, or conflicts between private land use and government resource management, or how land use practices influence deforestation and forest succession, an accurate position for each individual and their property is necessary. This is especially important in studies that need to locate plots of land on aerial photographs or satellite imagery. GPS data allow the researcher to explicitly identify the limits of each individual's property, which can then be used to analyze their behavior in relation to the landscape and the land cover within that piece of land. Even the ecotourism field is using GPS to aid in understanding, maintaining, and improving the natural resources of different areas.

BOX 2.4

Examples of Environmental Research Utilizing GPS Technology

In Jamaica, GPS was used to improve environmental management. Rybaczuk (2001) collected GPS data at environmentally important sites, such as coral reef diving buoys, to map their locations and improve environmental awareness and protection. Ganskopp (2001) used GPS collars to monitor cattle behaviors in order to identify patterns of overgrazing and erosion. Donohue et al. (2001) used GPS to collect data while searching for fishing gear debris off the Hawaiian Islands. Line data were collected with the GPS representing the search transects, while GPS point data were collected at all locations of debris. This allowed the researchers to identify high-density areas of debris that endanger coral reefs.

Link to GIS

A geographic information system is a "set of tools for analyzing spatial data" (Clarke, 1999, pp. 3). A GIS is first and foremost a software package that has the capability to display spatially referenced data, integrate data from various sources, analyze the data in a spatial database, and generate output in the form of a map. But there is more to a GIS than software. In order for a GIS to work properly, it must contain four other components: hardware, data, people, and methods. The GIS allows spatial phenomena to be converted into data, stored in a common format, retrieved at will, analyzed, and displayed (Clarke, 1995).

As early as 1990, it was apparent that the global positioning system would be extremely useful for collecting data for use in a GIS (Lang and Speed, 1990). The acceptance and use of GPS subsequently grew during the 1990s (Heywood et al., 1998), and it continues to grow. The incorporation of GPS into a social science project adds more than simply the x and y coordinates of a location. GPS data, when put into a GIS, give the researcher the ability to link aspatial data to real world coordinates. For example, a traditional survey of health clinics may not inherently contain any information pertaining to a clinic's location, with the possible exception of a street address. However, if a GPS point is collected at the clinic, then the survey data for the clinic can be linked to that point, resulting in a highly accurate location for that information. Once the researcher has a location for all of the re-

BOX 2.5

Examples of Population–Environment Research that Utilize GPS Technology

Fox et al. (1996) examined the impact of traditional cattle grazing on red panda (*Ailurus fulgens*) habitat in a portion of Langtang National Park, Nepal. A GPS was used to locate temporary grazing shelters and pastures in addition to collecting ground cover types within the study area. Pasture locations were mapped and GIS modeling was used to predict the most heavily grazed land. Land cover GPS data were used to classify a Landsat Thematic Mapper satellite image and map the land cover in the study area. Frizzelle and McGregor (1999) discuss the integration of GPS and GIS in facilitating social survey data collection for population–environment research in the Ecuadorian Amazon. GPS data is incorporated with satellite imagery and GIS-derived cadastral data to locate sample farms for interviewers to visit. The interviewers and farmers then used the resultant Survey Instrument Image Maps as a visual aid for creating sketch maps of land use on each farm. In Brazil, McCracken et al. (1999) use GPS to test the accuracy of and update a property grid. This GIS layer of farm plots is intended for integration with remotely sensed data in a GIS to better understand changing patterns of land use in conjunction with demographic data. Moran et al. (2000) used GPS to georeference satellite data, record household locations, and correct cadastral farm property data in Brazil. Priskin (2001) used GPS to locate nature-based tourism sites in Western Australia, and created maps that showed the distribution of various characteristics of those sites.

spondents, certain spatial analytical techniques can be applied to provide an extra dimension of analysis to the work. Some of the techniques for utilizing this "linked" data will be discussed in Chapter 13.

It is this ability to link aspatial data to spatial data, as well as the ability to integrate data from multiple sources, and then subsequently perform spatial analyses on that data, that is the true power of GIS. Given that GPS data are collected and stored in a digital format, it is a simple matter to add them as geographic features in a GIS. Some receivers output the data in a specific format that requires some manipulation before it can be entered into a GIS, while other receivers output the data in standard GIS formats (Heywood et al., 1998). The next step is entering the associated attribute data into a database program and importing the database file into the GIS, after which the attributes can be easily linked to their locations. However, some new GPS receivers allow attributes to be added to the data as it's being collected in the field (see the discussion of data dictionaries in Chapter 9), speeding up the attribution process and eliminating a separate attribute data entry, import, and linking process (Heywood et al., 1998). The flexibility of GPS makes it a valuable tool for collecting GIS data.

Comparison of GPS to Other Methods of Locating Phenomena

The use of GPS has begun to replace older methods of locating phenomena in the real world. Satellite-based technologies such as GPS, which are used for GIS data collection, have "led to significant advances in the speed and accuracy of traditional ground survey methods" (Robinson et al., 1995). Prior to widespread availability of GPS technology, researchers wanting to incorporate a geographic location with their data had to resort to methods that were less accurate, more time consuming, and/or very expensive. Problems and obstacles associated with implementing these methods often led to low accuracy in the coordinates or, in the worst case, the researcher being unable to incorporate the spatial perspective into the study. Four such methods are discussed here: approximating locations with hardcopy maps, traditional surveying, digitizing, and address geocoding.

Using maps to approximate coordinates

Researchers have used hardcopy maps to locate phenomena for centuries. This approach is the oldest of the four discussed in this chapter. There are two different ways that one can use a map to locate objects of interest, and both are contingent on the map having a well-defined coordinate system grid, or *graticule*. The graticule is common to many reference maps and to all topographic maps, and is basically a crisscross of horizontal and vertical lines that have defined coordinates and are separated by a regular interval. The simplest way to determine the coordinates of a map feature is to eyeball the location. That is, the map reader finds the feature on the map and using the coordinate grid, approximates the coordinates. This is a quick-and-dirty method, which provides a general idea of the location of the feature, but is not highly accurate.

The second, more accurate way of approximating coordinates from a hardcopy map is to use a map ruler or map protractor. These are similar to standard drafting tools, but have distances based on common map scales. The tool is lined up with the coordinate grid line nearest to the feature, and provides a more accurate measure of the feature's distance from those known coordinates. Still, even with the most detailed map ruler, the highest level of accuracy that can be achieved is dependent on the map scale (see Table 2.1), which is easily calculated, and user error, which is of unknown magnitude. The condition of the map can also affect the calculated accuracy, as wrinkles and creases in maps can skew distance measurements.

With either method, each feature has to be measured and the coordinates recorded and later entered into a database program or directly into the GIS. This is a time-consuming process and can be extremely tedious if one is attempting this with many features.

Traditional surveying

Traditional surveying is a means of manually locating the precise coordinates of a location using a known reference base (i.e. a horizontal and vertical control network), knowledge of geodesy (see Chapter 4), and measured angles and distances to perform the trigonometric calculations (Robinson et al., 1995). These techniques are well established, and have been in operation for a long time. The benefit of manual surveying is that the resultant coordinates are extremely accurate. However, that advantage is countered by the high costs of a large amount of equipment and a crew of at least two people (one person can work alone, although this requires the use of even more expensive, cutting-edge equipment), the fact that this approach takes a considerably longer time to complete for

Table 2.1 Scale-dependent errors, based on the US national map accuracy standards

Map scale	Expected error
1:1000	0.85 m
1:5000	4.23 m
1:10,000	8.47 m
1:25,000	12.7 m
1:100,000	50.8 m
1:250,000	127.0 m
1:1,000,000	508.0 m

one location, and dependence on the existence of a known reference base. In reality, many areas of the world have no reference control networks established, precluding the use of traditional surveying methods. GPS, in contrast can be used anywhere on the planet. Traditional surveying is a good method if one has a reference base, the proper equipment is available, and there are only a few locations for which one needs the coordinates. However, it's not so efficient for a large number of locations in comparison to GPS data collection.

Being a manual data collection technique, the coordinates must be recorded by hand and later entered into a database for eventual input into a GIS, if so desired. However, GPS is an electronic data collection method that stores the calculated coordinates digitally, which simplifies the process of getting the data into a GIS, but it can be less accurate than traditional surveying depending on the GPS equipment used and the corrections applied to the GPS data.

Digitizing coordinates

A third approach is based on the approximation of phenomena with hardcopy maps, but incorporates GIS and increases accuracy of location. This method is known as *digitizing*. In digitizing, a hardcopy map is placed on a large flat tablet that is connected to a computer. The data entry tool is a handheld puck with crosshairs and a keypad that is connected to the tablet, which has a dense network of electronic sensors beneath the surface that can sense the location of the crosshairs at all times. To digitize a point, the crosshairs are placed over the point and one location is entered into the computer using the keypad. To digitize a line on a map, the user traces the line with the crosshairs and continuously enters points along the length of the line. Each location is entered with more accurate coordinates than can be obtained using manual techniques. The crosshairs of the digitizing puck allow for greater accuracy than a map ruler, thus substantially reducing the user error and leaving the digitizing tablet coordinate transformation, map quality, and map scale (see Table 2.1) as the main error sources. The user error described in manual map measurement is eliminated, but a new user error is created. This error source is the ability of the user to accurately trace the shape of the lines that are being digitized. The more skillful user will be able to digitize a truer representation of the line than a novice, however even the steadiest hand is prone to mistakes.

Digitizing is also a faster method of collecting feature coordinates than map approximation, but is more expensive due to the necessity of purchasing a digitizing tablet and related hardware, software, and drivers. In addition, the coordinates are entered in a digital format, thus eliminating the intermediate steps of hand-recording coordinates and entering them into a database, both of which are subject to transcription errors. Digitizing makes the transition to a GIS data set much smoother.

Address geocoding

The fourth, and most recently developed, method of locating people and places is called *address geocoding*. It is much different in that instead of using hardcopy maps or survey equipment, it uses roads and addresses within a GIS to obtain a relative location for people and places. Address geocoding requires that the researcher have a data set containing a road network, and every road segment in the network must contain the road name and the address range. For instance, a particular block of Main St may contain the addresses ranging from 100 Main St to 200 Main St. The GIS uses this attribution to locate addresses. First, it locates all of the road segments that contain the same name as the address of interest. Second, it selects the road segment along that road with an address range that contains the address number of interest. Finally, simple geometry is used to locate the address along the segment. For example, a researcher wants to find the location of 150 Main St. The GIS locates all road segments that comprise Main St, and then finds the segment with the address range in which 150 falls–in this case, the segment with the range 100 to 200. Since the number 150 is halfway between the two extremes of the address range, the resultant location would be a point on the road midway between the two ends of the segment. It's important to note, however, that the true location may not be in the middle of the road segment, as addresses are not always evenly spaced along every street segment.

Benefits of GPS over these methods

First, it should be quite apparent that GPS is a more accurate method of data collection than map approximation and digitizing. The inherent error of GPS (see Table 5.1) is rarely greater than the map scale error, and even when it is greater, using a GPS unit to collect data at the location ensures that the true, absolute location is captured. While map approximation and digitizing may be less time consuming and less expensive than GPS-based fieldwork, especially if the field site is far away, the resultant accuracy far outweighs those negatives (assuming that a higher level of accuracy is desired and funding is available).

In comparison to traditional surveying, GPS is a much faster approach. It is less expensive than traditional surveying if recreational or mapping grade GPS receivers are used (see Chapter 5), but is also less accurate. However, if extremely high accuracy is necessary, there are survey quality receivers available at higher costs that give the same level of accuracy (sub-centimeter) as traditional surveying and still provide a timesaving. GPS can be used anywhere in the world and is not dependent on the existence of a known reference base or control network, unlike traditional survey methods.

The use of GPS for locating houses, businesses, or other places has a definite benefit over address geocoding. First of all, GPS allows the user to record an *absolute* coordinate pair for the location of the building, whereas address geocoding creates a point on the road segment at a *relative* location (see Figure 2.1). Second, GPS can be used in areas where the address infrastructure is not well developed. In the United States, the usefulness of address geocoding tends to break down in rural areas, as the GIS cannot effectively handle rural route addresses. And in areas of developing countries where there may be no address infrastructure or even roads for that matter, address geocoding is ineffectual. By using GPS, the researcher may circumvent these potential problems.

However, there are some cons when comparing GPS to address geocoding. GPS is a more expensive and time-consuming approach. A trained analyst can sit at a computer and, with the proper data and software, address geocode hundreds of locations in minutes. In comparison, collecting GPS coordinates of hundreds of locations may take days or weeks, and cost hundreds or thousands of dollars. So when making the decision of which method to use, the factors to consider are cost, time, infrastructure, and overall accuracy requirements.

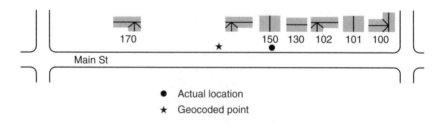

Figure 2.1. A comparison of a geocoded address and absolute GPS location

Summary

The global positioning system is a wonderful resource for collecting geospatial data in the field. The technical prowess of the GPS receivers continues to improve, and as it does so the cost continues to decrease, making it an affordable tool for any research project interested in understanding the influence of geography on data. There is a relatively short learning curve to understand the operation of the system, use a receiver, plan and implement data collection in the field, and utilize the data after completing fieldwork. All of these steps are discussed in great detail in later chapters and reinforce the importance of the global positioning system as a valuable and worthwhile investment for incorporating the spatial perspective into any research project.

3
What is GPS?

Before incorporating GPS into a research project it is important to understand the workings of the system. Even though the receivers themselves are usually quite simple to operate, there is a lot going on behind the scenes. A basic understanding of how the system was developed, the individual components and how they interact to calculate a position can be very valuable.

History

The development of GPS has its roots in the United States Department of Defense. As far back as the 1960s, the US Navy and Air Force developed satellite based radiopositioning systems to provide highly accurate positioning and navigation support for submarine based ballistic missiles and other purposes (US Department of Defense, 1991).

In 1973, the US Department of Defense combined the Navy and Air Force systems to create one comprehensive system to provide accurate data on position, velocity, and time to both military and civilian users. It was given the name "NAVSTAR" and placed under the auspices of a joint program office, NAVSTAR JPO, which had representatives from all of the US military services.

NAVSTAR JPO was responsible for maintenance, updates, and operation of the GPS components until 1996 when the Interagency GPS Executive Board, or IGEB, was given responsibility over the system. According to the directive issued by President Bill Clinton, IGEB serves the needs of both the US military as well as the civilian community. The board is co-chaired by representatives from the US Department of Defense and Department of Transportation. The US Departments of State and Commerce are also principal members, with other US government agencies also serving on the board (IGEB, 1997).

Role of civilian and international users in GPS

Despite the fact that GPS was developed by the US military, civilian access to the signal is important and is formally guaranteed by an executive order issued by President Ronald Reagan and reasserted by President Clinton in his directive of 1996 (IGEB, 1997; US Department of Defense, 1991). Accordingly, civilian users are represented on the IGEB, and the representative from the US Department of Transportation is responsible for the oversight of the civilian policy and related issues.

Even though GPS and the executive board are US-based, international users are considered an important user group. The responsibility of representing international users on the executive board has been given to the US Department of State.

Three Components of GPS

The global positioning system is the integration of three main components: *space*, or the satellites orbiting the earth; *control*, the infrastructure monitoring and operating the satellites; and the *users*.

Space

As the name suggests, the space segment consists of the satellites orbiting the earth. The fully operational system has 28 satellites orbiting 12,500 mi (20,110 km) above the earth. The constellation of orbits (Figure 3.1) was designed so that at least four satellites are visible anywhere on Earth at any time (US Department of Defense, 1991). Each satellite broadcasts radio signals that receivers can use to calculate a position.

It is beyond the scope of this book to describe in detail the multiple components of the signal. However, it is important to point out that the satellites send out radio signals that contain a considerable amount of data. Information on satellite health, satellite position, as well as data that can be used to determine the satellite time are all transmitted via these signals.

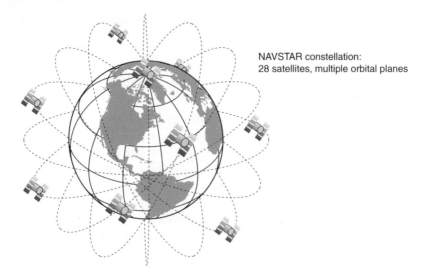

NAVSTAR constellation:
28 satellites, multiple orbital planes

Figure 3.1. The GPS satellite constellation consists of 28 satellites orbiting 12,500 mi above the Earth in orbital planes designed to keep at least four satellites above the horizon anywhere on the planet

Control

Since the integrity of the system relies on the satellites precisely maintaining their orbit, it is imperative that their positions in space be monitored. This is the responsibility of the GPS Master Control Station (MCS) in Colorado Springs, Colorado. The Master Control Station and its five associated monitoring stations around the world monitor the position of the satellites in their orbits, the health of each satellite, and the signals they transmit (US Department of Commerce, 1998). The Master Control Station is staffed and operated by the US Air Force 24 hours a day to ensure the system functions properly. Personnel at the Master Control Station will upload new time and orbital data to each satellite on a regular basis. Included in this transmission is an *almanac*, which contains satellite orbital positions, satellite status, clock corrections, and atmospheric delay parameters. Also included in the uploaded transmission is *ephemeris* data, which contains the predicted positions of satellites. The almanac and ephemeris data reduce the error present in the signal by resetting the errors in time and position that have gradually accumulated since the last update. Therefore the role of the personnel of the MCS is crucial in maintaining the accuracy of the system (US Department of Defense, 1991).

Users

Of course, the third segment is made up of the users of the GPS signals. In addition to social science researchers, the aviation industry, transportation, agriculture, consumers, public service sector, and many others rely on the system.

No matter the application, all users of GPS take the ephemeris and almanac transmitted by the satellites and use it to derive new information: time, position, and/or velocity. This derived information allows users to answer basic questions such as "Where am I?," and "What time is it?" with a level of accuracy that was unthinkable prior to GPS. This accuracy has led to some innovative uses of GPS:

- aviation – direct routing of airplanes resulting in fuel savings, and closer aircraft separation standards;
- communications – high precision timing for network security protocols and validation of information transmission;
- ground transportation – fleet monitoring and dispatch;
- public safety – emergency vehicle routing and tracking;
- shipping – cargo inventory management and tracking; search and rescue services;
- entertainment – time signals sent by GPS satellites are used to coordinate national broadcast commercial breaks and program length; special effects timing for film and video productions (US Department of Commerce, 1998).

The wide range of commercial, government, military, and recreation uses for GPS has resulted in a tremendous growth in the demand for GPS receivers and ancillary equipment. A 1997 study by the consulting firm The Freedonia Group, cited in a US Department of Commerce publication, estimated that the worldwide GPS market would grow from US$1.5 billion in 1996 to reach US$6.5 billion in 2001 and achieve US$16.4 billion by 2006. The same Department of Commerce report estimated that in the year 2000 the GPS user base would grow by 2 million users per month, with a potential 40 million users in the US consumer and recreation market alone (US Department of Commerce, 1998). Clearly, GPS is vital part of the twenty-first-century information infrastructure.

How Does it Work?

Given the complicated and expensive technology involved, the principles behind the system are amazingly simple. The basic principle relies on the familiar equation: *distance = rate × time*. By knowing how fast something is traveling and how long it takes to arrive, the distance it travels can be calculated using simple mathematics. In the case of GPS, the object traveling is a radio signal, which travels at the speed of light. By receiving multiple signals from multiple satellites, a process similar to triangulation can be used to determine the exact position of the user. To illustrate this, a simplified description of the process of calculating a position from the signal, taken from The *NAVSTAR GPS User's Overview* published by the United States Department of Defense in 1991 will be presented.

Imagine a ship in a foggy bay. Its captain does not know the ship's precise location but the captain knows that there are foghorns at three lighthouses each with unique sounding horns that are regularly sounded once every minute. On board the ship, the captain has a very precise clock that, for our purposes, we will assume is synchronized with the lighthouses.

At 10 seconds past the minute mark, the captain hears the signal from the first foghorn. Since the captain knows that the speed of sound is 340 m per sec, he can use the *distance = rate × time* equation to calculate that his ship is 3,400 m from the lighthouse (Figure 3.2a). Furthermore, the captain also has a chart that shows the exact location of the lighthouses, and is able to draw a circle with a radius of 3,400 m around the lighthouse. The perimeter of this circle, known as a range, represents all possible locations of the ship. Using the same procedures for the second lighthouse, the captain is able to plot the intersection of the two ranges (Figure 3.2b). The ship's position is narrowed to where the two circles intersect. Given there are two possible locations, the captain must rely on a third foghorn to help get a more precise location. By calculating the distance from the third foghorn the captain knows his position is at the intersection of the three circles (Figure 3.2c).

Effect of timing errors

Let's say that the captain's chronometer is off by 1 sec. In other words the sound from the foghorn seems to arrive 11 secs after the minute, instead of 10 secs after the minute. This causes the captain to calculate he is 3,740 m away from the foghorn (340 m/sec × 11 secs). This same 1 sec error is replicated with foghorn two. The intersection of foghorn one's range and

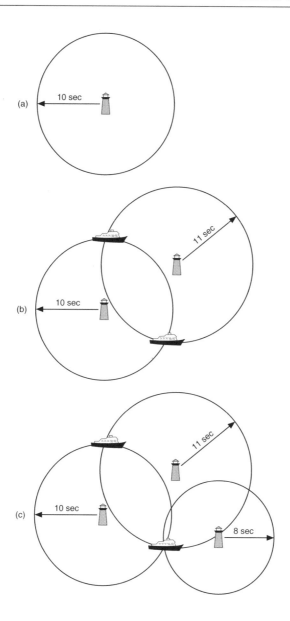

Figure 3.2. The captain determines the ship's location by measuring distances (based on time) from three lighthouse foghorns. (a) One measurement puts the ship somewhere on a circle. (b) Two measurements reduce the possible locations to two points where the circles intersect. (c) Three measurements determine a single location

foghorn two's range will be be incorrect (see Figure 3.3a). Once again, by using a third foghorn this error can be eliminated. The apparent intersection of the ranges for foghorns one and three, as well as the apparent intersection of the ranges of foghorns two and three are actually incorrect due to the chronometer's inaccuracy. Because of this error, there is no common intersection for all three foghorns. The captain can then adjust his chronometer until all of the foghorn ranges converge at one location (Figure 3.3b).

With GPS, the satellites are the foghorns at the known locations and the GPS receiver replaces the captain's chronometer and navigational chart. The satellites broadcast radio signals that consist of a set of ones and zeros corresponding to a predefined pattern, unique to each satellite. Encoded in this signal is information on the satellite's offset from the GPS Master Clock, known as clock bias. Also included is information on the satellite's overall health, high precision ephemeris data and an almanac that describes the position of other satellites.

The GPS receiver is able to use this signal to determine which satellite is sending the signal and its location, and if the receiver's clock was perfectly synchronized with the satellite's clock, it could then calculate a range or position. However, since each satellite contains atomic clocks accurate to a nanosecond, it is not feasible to perfectly synchronize it and the receiver. As a result, there is likely to be clock bias present in the receivers. To solve this, the receivers have their own internal clock and mathematically adjust it to resolve the clock bias, much like the captain did in the example provided above. Once the GPS receiver has a signal from 3 satellites, it can calculate a 2-D position (x,y); when 4 or more satellites are available a 3-D position (x,y,z) can be calculated. Theoretically, the resulting position should be correct within the parameters of any clock bias present. However, as will be explained in Chapter 5, there are a variety of potential errors that can result in the position being off by tens of meters or more.

Summary

The principles behind GPS are deceptively simple, yet it requires an enormous investment in technology and infrastructure for it to work reliably. For GPS users it may not be necessary to fully understand the minutiae about how the system works, but it is important to have a basic understanding of the technology going on behind the scenes. This keeps GPS from becoming a mysterious black box. Users who understand the basics of the system can then more effectively troubleshoot problems and can collect better data.

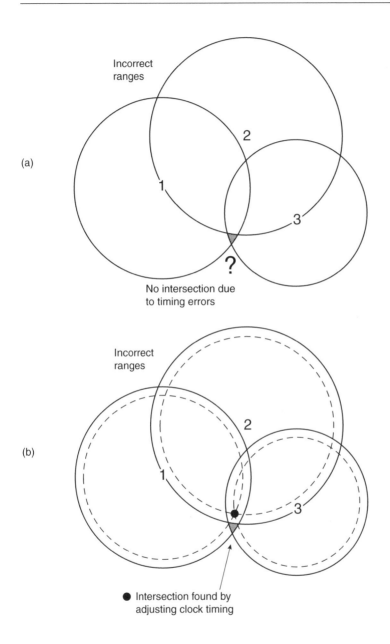

Figure 3.3. (a) The ranges from three incorrect clocks won't intersect at a single point. (b) The error can be removed by adjusting the clock timing until the three circles converge on a single point

4
Coordinate Systems and Datums

The previous chapter gave an overview of how the global positioning system works, a basic understanding of which is necessary to ensure a successful GPS data collection effort. It is also necessary to understand coordinates and coordinate systems. Coordinates are, after all, the primary output of the global positioning system. The locational information recorded with a GPS (or displayed on a map, or stored in a GIS) can be expressed in any of a number of defined coordinate systems, all of which – at their simplest – are much like the coordinates in simple graphing exercises with which a high school algebra student is familiar. These systems allow us to pinpoint a specific location on the earth's surface much as that student might use a pair of coordinates (x,y) to define a point on a plane, or three coordinates (x,y,z) to locate a point in a three-dimensional space.

However simple that may seem, the reality of defining positions on the earth using a system of coordinates is actually much more complex. In fact, there is an entire field of study known as *geodesy* that is devoted to measuring the planet's size and shape and developing systems for precisely determining and specifying locations on it. This chapter just barely scratches the surface of these topics – just enough for a social scientist user to understand the coordinate data collected by a GPS and avoid making some very fundamental mistakes. It may seem complicated at first, but understanding the material in this chapter will save many headaches later.

The earth is not a sphere, of course. It has a very complex shape and accurately referencing a location on this complicated shape requires that simple coordinates be teamed with several other pieces of information. One such item is an *ellipsoid* (one of several models of the shape of the earth), a geometric shape that is used as the basis for *geodetic coordinates*, a three-dimensional coordinate system. Another is a *datum*, something that defines how this ellipsoidal model of the earth's shape corresponds to its actual shape and size. And finally, there is a *map projection* that lets us reference locations in two dimensions (like on a map) using a *planar coordinate system* instead of the three-dimensional geodetic coordinates associated with the ellipsoidal model of the earth. This chapter explains how these elements – ellipsoid, datum, geodetic coordinates, map projections, and planar coordinates – are interlinked and why they are important to a social science GPS user.

One point to emphasize is that there are many combinations of these elements that are used to reference positions on the earth, and this variety is reflected in the configuration options of most GPS receivers. Most notably, the global positioning system itself makes use of an entirely different set than those used to make more traditional maps or found in most GIS databases. Since most researchers using GPS technology are collecting data that will be used in conjunction with other mapped information, whether within a GIS or even just to compare with other locations on printed maps, it is important to understand how these different coordinate reference frameworks relate to one another.

Geodetic Coordinates and Datums

Approximating the earth's shape

In order to pinpoint a location on the earth's surface, we have to have some representation, or model, of that surface to which we can apply a coordinate system. The surface of the earth itself is far too complex a shape to use as the basis for a coordinate system – we must use a simplification. The simplest approximation of the earth's shape is a sphere with a radius of 6,371 km. But since the earth is slightly flattened at the poles and wider at the equator, this spherical model deviates enough from the shape of the earth to make it unsuitable for mapping projects in which high accuracy is a significant concern. Most GPS users are likely to be working with mapped information at a larger scale and with much more stringent accuracy requirements than are achievable using a spherical model.

Instead we represent the earth's shape as an ellipsoid, which can be visualized as an ellipse rotated around its shorter axis to form a three-dimensional shape, with its polar axis slightly shorter than its equatorial axis. This more accurately captures the earth's polar flattening and equatorial bulge than the spherical model. The size and shape of the ellipsoid are specified by the length of its semimajor axis (half of the horizontal, or equatorial, axis) and by the amount it is flattened (how much it deviates from a sphere).

BOX 4.1

Geodetic vs. Geographic Coordinates

Geodetic coordinates: Latitude and longitude on an ellipsoidal model of the Earth. These are the basis for all other mapping coordinate systems. For example, the UTM coordinate system described later in this chapter is based on a projection from geodetic coordinates (three-dimensional, on the ellipsoid) to a two dimensional, planar coordinate system.

Geographic coordinates: This term is frequently substituted for geodetic coordinates, but this is potentially confusing because many people use geographic coordinates in reference to latitude and longitude on a *spherical* model of the Earth. We strictly refer to geodetic coordinates in this book.

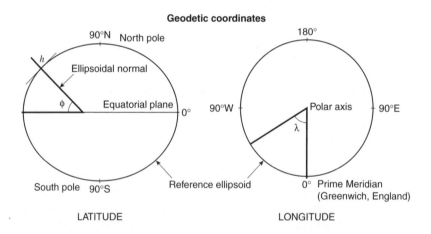

Geodetic coordinates

LATITUDE   LONGITUDE

ϕ latitude: the angle between the equatorial plane and the ellipsoidal normal

h height above or below the ellipsoid (a measurement of elevation)

λ longitude: the angle east or west of the Prime Meridian, an imaginary line passing through Greenwich, England

Figure 4.1. Geodetic latitudes and longitudes specify locations on a reference ellipsoid

Locations on the ellipsoid are referenced using *geodetic coordinates*, expressed in degrees of *latitude* and *longitude* (Figure 4.1). The origin of this system is located at the center of the sphere, with latitude (ϕ) defined as the angle between the equatorial plane and the ellipsoidal normal, a line perpendicular to the surface of the ellipsoid at the location being referenced. Longitude (λ) is the angle east or west of an imaginary line known as the Prime Meridian, which passes through Greenwich, England. Height (h) indicates relative position above or below the ellipsoid.

So the ellipsoidal model satisfies our criteria to be the basis of a coordinate system. It is a reasonably good (though far from perfect) approximation of the earth's shape, and it is very simple to define mathematically,

---

BOX 4.2

Latitude and Longitude Coordinates

Latitude and longitude are angular coordinates expressed in degrees. One degree is divided into 60 arc *minutes*, and one minute is further divided into 60 arc *seconds*.

Positive latitudes are north of the equator, and negative latitudes are south. Positive longitudes are east of the Prime Meridian (at Greenwich, England), and negative longitudes are west.

Working in degrees, minutes, and seconds is cumbersome. Most mapping packages expect latitude and longitude to be expressed in decimal degrees, or very occasionally, in decimal minutes.

Decimal Minutes = Degrees, Minutes + (Seconds / 60)

Decimal Degrees = Degrees + (Minutes / 60) + (Seconds / 3600)

The following latitude/longitude positions are equivalent:

Degrees, Minutes, Seconds
latitude: 39° 45′ 30″
longitude: -82° 30′ 00″

Decimal Minutes
latitude: 39° 45.5′
longitude: -82° 30.0″

Decimal Degrees
latitude: 39.758′
longitude: -82.500′

BOX 4.3

Ellipsoid vs. Spheroid

Note that the term *spheroid* is sometimes used instead of *ellipsoid*. Both are correct, although spheroid is more precise. Ellipsoid is more commonly seen in GPS documentation, so it is used here.

making it easy to pinpoint a location on it with only two angular measurements: the geodetic latitude and longitude. An ellipsoid used for this purpose is known as a *reference ellipsoid*. Many reference ellipsoids of slightly varying dimensions and degree of flattening have been defined over the years in attempts to best approximate the size and shape of the earth.

Geodetic datums

While an ellipsoid is a convenient abstraction useful for defining a coordinate system, specifying a latitude/longitude coordinate pair on a reference ellipsoid does not itself define a location on the earth. We have to link the reference ellipsoid (and the coordinates defined on it) to actual locations on the earth's surface. That is, the position of a chosen reference ellipsoid must be fixed relative to the earth. This is accomplished by the definition of a *geodetic datum.*

Historically this could be done in a manner that would only yield accurate positions for a limited area, so all geodetic datums were by necessity local, or regional, in extent. To define a local geodetic datum, a physical location on the earth is assigned to a point on the ellipsoid identified by its geodetic latitude and longitude (Figure 4.2). This is the origin of the datum. The orientation of the ellipsoid relative to this location is chosen so that the ellipsoid best matches the area for which the local datum is to be applicable.

This location linking the origin of the datum at a physical location on the earth with a point on the ellipsoid – in effect where the earth and the ellipsoid are "attached" to one another – is marked with a small monument sunk into the ground or somehow else made permanent and immovable. This linkage *realizes* the datum, associating an absolute location with a point on the abstract surface of the ellipsoid. Surveyors then link this monument at the datum's origin to monuments at other locations with coordinates specified on the ellipsoid, forming a network of geodetic control points that can be used as the locational framework for making maps. Defining a geodetic

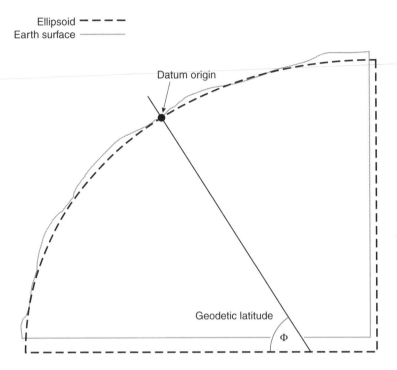

Figure 4.2. A simplified representation of a local geodetic datum. The datum provides a link between geodetic coordinates on the ellipsoid and actual locations on the Earth. (Differences between the ellipsoid and the Earth's surface are exaggerated)

datum in this manner allows us to describe locations on the complicated and irregular surface of the earth using coordinates that are based on the simplified and regular geometry of an ellipsoid.

There are two points to emphasize here:

1   knowledge of the underlying geodetic datum is necessary to specify an absolute location. A latitude and longitude coordinate pair does not suffice;

2   mapped coordinate data collected or displayed using one geodetic datum cannot be directly used in conjunction with mapped data based on another. The same latitude and longitude coordinate pair will appear in different locations on the two maps. One set of the data must be transformed to the datum of the other.

Over time, many local datums and their associated ellipsoids have been defined for mapping different regions of the globe. During the course of the twentieth century, most countries developed geodetic systems incorporating regionally appropriate local datums. Readers familiar with US and Canadian topographic maps may recognize the North American Datum of 1927 (NAD27), based on the Clarke 1866 ellipsoid, and its replacement, the North American Datum of 1983 (NAD83), based on the Geodetic Reference System 1980 ellipsoid (GRS80). The European Datum of 1950, with its origin in Potsdam, Germany, became the basis for the national map series in many European countries.

Elevation and vertical geodetic datums

The discussion so far has focused on datums used to define horizontal positions – in geodetic coordinate terms, positions east or west of the Greenwich meridian and north or south of the equator. Vertical positions, or elevations, are usually expressed relative to mean sea level, which is an abstract surface used as a reference, just as the ellipsoid is in horizontal positioning.

Another model of the shape of earth is based on its gravity field. This model is the *geoid*, the shape of which is defined to be everywhere perpendicular to the direction of gravity (i.e. a plumb line is always perpendicular to the geoid), and the force of gravity is equal to the force of gravity at mean sea level (Iliffe, 2000). If the earth were made completely of water, and there were no forces such as currents or tides affecting its shape, then the geoid would be a perfect ellipsoid corresponding exactly to sea level. Of course this is not the case, instead variations in the internal composition and density of the earth cause the geoid to dip or rise in places, making it an irregular shape. It doesn't match mean sea level exactly, though it is very close, nor does it correspond to the topographic surface of the earth.

Traditionally, elevations are measured using a process called leveling, in which heights at different points are measured relative to one another using surveying equipment that is kept level at each point using the direction of gravity (perpendicular to the surface of the geoid) as a reference. These heights are measured relative to known elevations in vertical geodetic control networks that use mean sea level as their zero point. *Vertical geodetic datums* define the relationship between mean sea level in a region and the elevations in the vertical control networks.

Nearly all maps express elevation relative to mean sea level. The geoid and mean sea level coincide almost perfectly, differing from one another by less than 1 m globally, so elevation above mean sea level is sometimes referred to as *height above the geoid*, and for all practical purposes they are the same. Either way, a variety of vertical datums are in use for mapping purposes, and just as with horizontal positioning, vertical positions are only meaningful in the context of the particular vertical datum used as the reference to specify them.

WGS84 and the global positioning system

In the latter half of the twentieth century, advances in surveying and geodetic measurement technologies both allowed and necessitated the development of a datum appropriate for global use. Satellite technologies for geodesy and mapping required that the center of the ellipsoid and the center of the geoid coincide in the datum, meaning that the ellipsoid is *geocentric* around the center of the earth's gravity, just as the orbits of the GPS satellites are. The United States Department of Defense undertook the development of a global, geocentric datum, a process that culminated in the World Geodetic System of 1984 (WGS84).

The global positioning system uses the WGS84 datum. Several enhancements for increased accuracy were made to it during the 1990s, but it is still known by the same name. The GRS80 ellipsoid was originally adopted for the datum, but there were minute differences between it and the specifications of the ellipsoid actually incorporated into WGS84, which is therefore more precisely referred to as the "WGS84 ellipsoid" (Slayter and Malys, 1997). Confusingly, references to both the GRS80 and WGS84 ellipsoids can be found in the documentation accompanying different GPS receivers. However, because the differences between the two are so small, they are identical from the standpoint of the GPS user and this discrepancy can be ignored.

GPS receivers use the WGS84 datum to perform the positioning calculations and to store the resulting positions. By default, most units also report WGS84 positions on their displays. Since most mapped information available today has been developed based on regional datums as described above, it is crucial that GPS users are aware that until they change the settings, positions reported by their GPS receivers will not match positions taken from other maps or GIS data for their project, if another datum was used. However, the parameters for transformation between WGS84 and many regional datums are well known and GPS receivers are generally capable of performing them. For the user it is simply a matter of configuring the GPS receiver to report coordinates using the datum appropriate for their project.

GPS-derived vertical positions are handled differently as well. The global positioning system calculates and expresses elevations as height above the WGS84 ellipsoid, not in relation to a mean sea level datum, as is the norm. If a project requires the collection of elevation information, then researchers should verify that their GPS receivers or post-processing software can convert elevations to height above the geoid (a reasonable sea level approximation, as described above) or even to specific mean sea level datums.

Planar Coordinate Systems

Projecting coordinates

Geodetic coordinates (latitude and longitude on an ellipsoidal model of the earth) are three-dimensional, but maps and computer screens are flat, necessitating that mapped information be displayed in two dimensions. Furthermore, working with geodetic coordinates is mathematically and computationally difficult compared to working with coordinates in two dimensions. Making measurements on maps and developing GIS software to perform analyses are much easier if the coordinates are arranged in a two-dimensional, rectangular Cartesian coordinate system with x and y axes, often referred to as a *planar* coordinate system. For example, measuring the distance between two points using two-dimensional Cartesian coordinates is a simple application of the Pythagorean theorem–much simpler than the ellipsoidal trigonometry required with geodetic coordinates to make the same measurement.

A *map projection* is used to transform mapped information from three to two dimensions, by "projecting" it from the ellipsoid to a flat surface (Figure 4.3). Projections can be carried out in a variety of ways, usually involving mathematical formulas to reduce the three dimensions to two. Whatever the method, projecting from the ellipsoid to a flat surface can not be done without introducing some distortion into the resulting map. (Try to imagine removing an orange peel in one piece, and then trying to flatten it. Making it perfectly flat would require either splitting or stretching the peel in places to smooth it out.)

They can be loosely categorized into three families based upon the idea of a *developable surface*: a plane, cone, or cylinder onto which features on the earth are projected (Figure 4.4). To visualize this, imagine a sheet of paper placed against or wrapped around a translucent globe with a light bulb illuminating it from inside, projecting the shapes of land and oceans onto the paper. The shape or size of features appearing on the paper further

Three-dimensional   Two-dimensional (projected)

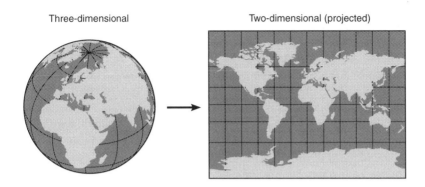

Figure 4.3. Map projections make it possible for us to map the three-dimensional world in two dimensions

away from where it touches the globe will be more distorted than those that are closer. Where the paper is tangent to the globe there is no distortion at all. If the paper is wrapped around the globe to form a cylinder or a cone, it can then be unrolled and flattened without having to stretch or tear it. Of course, in planar projections the developable surface is itself flat.

Projections are chosen by mapmakers in order to minimize the distortions for the particular area being mapped. The places where the developable surface intersects or is tangent to the ellipsoid are known as *standard points* or *standard lines*. Distortion increases with distance from a standard point or line, so designing or choosing a projection for a particular area is

Cylindrical   Planar   Conic

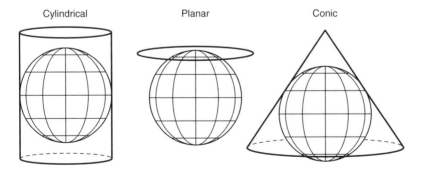

Figure 4.4. Cylindrical, planar, and conic developable surfaces

partly a matter of orienting the developable surface to maximize the area within a map that is near one of them. Projections are usually named for the cartographer who designed them, plus some descriptive information about the characteristics of the projection. Two examples: transverse Mercator (designed by Mercator and the cylindrical developable surface is oriented transversely), and Albers equal area (designed by Albers to avoid distortion in the sizes of areas shown).

Once the area being mapped is projected, a two-dimensional coordinate system is applied by laying down a rectangular grid, specifying the geodetic coordinates of the point on the grid that will be the origin (0,0), and choosing the distance units to be used, usually meters or feet (Figure 4.5). The coordinates along the x-axis (running east-west) are *eastings*, while the y-axis coordinates (running north – south) are *northings*.

Universal Transverse Mercator

Social scientists working in different parts of the world are likely to encounter a variety of different projected coordinate systems. However, the Universal Transverse Mercator (UTM) coordinate system is a special case in that it is designed to be usable worldwide and is widely used by many national and international organizations. Many maps produced by national

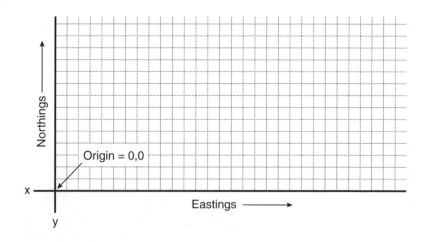

Figure 4.5. Projected, two-dimensional Cartesian coordinate systems are used for most mapping and GIS applications

---

BOX 4.4

Projected Coordinates

Geodetic coordinates are projected into two dimensions, generating projected, or planar, coordinates. The familiar *x* and *y* coordinates of a two-dimensional coordinate system are known as *eastings* (along the *x*-axis) and *northings* (along the *y*-axis) when used in mapping. Projected coordinates are usually expressed as an easting and northing coordinate pair. For example, this UTM coordinate location in Zone 17, North:

620,057.00 E   3,200,428.50 N

indicates a location 620,057.00 meters east of the false origin of the UTM zone, and 3,200,428.50 meters north of the equator.

---

mapping agencies display UTM grid coordinates in addition to those of the national grid on their maps. The UTM system divides the globe into sixty zones, each 6° of longitude wide and extending from 80°S to 84°N latitude (Figure 4.6). Each zone is projected using a transverse Mercator projection, a cylindrical projection with the cylinder oriented along a meridian running north–south in the center of the zone. The cylinder intersects the globe, so there are two standard lines on either side of the central meridian.

Each zone is further split by the equator into northern and southern sub-zones, with the intersection of the equator and the zone's central meridian used to define a separate origin for each. For the northern sub-zone, the central meridian is given an easting of 500,000 m and a northing of 0 m, locating its (0,0) origin on the equator and outside its eastern boundary. The central meridian for the southern sub-zone is again given an easting of 500,000 m, and the equator is given a northing of 10,000,000 m, locating its (0,0) origin to the south and west of the zone. Note that by specifying the origins in this manner, all coordinates within each zone are positive, a common characteristic of projected coordinate systems.

Thus the UTM system is actually 120 coordinate systems, which together span the globe, covering all but the polar areas. The transverse Mercator projection and its orientation within each zone result in a minimum of mapping distortion within and slightly beyond each zone's boundaries. Researchers using UTM should select the appropriate zone number for their study areas, which usually can be found on topographic maps of the area. Refer to

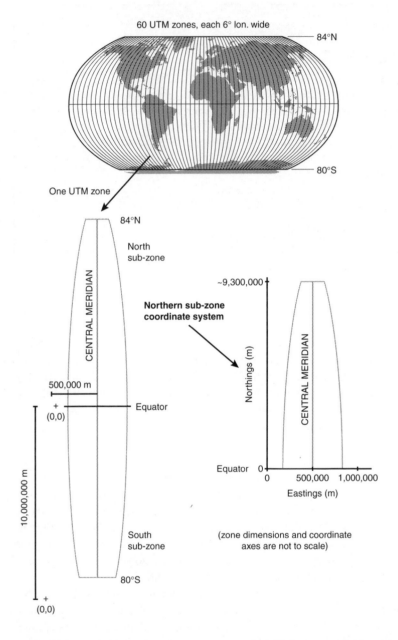

Figure 4.6. The Universal Transverse Mercator (UTM) system divides the globe into 60 zones. Within each zone there are two coordinate systems, one north of the Equator and one south

Appendix C for the latitude/longitude boundaries and central meridians of each. Study areas that cross UTM zone boundaries are acceptable, but as a general rule extending past the central meridians of the zones adjacent to the one in use should be avoided.

The geodetic datum appropriate for the project area is geographically dependent. Because the UTM zone system is used worldwide, it does not incorporate a specific datum. Local datums appropriate to the region being mapped are chosen to use with it, or a global datum like WGS84 is used.

National mapping grids

Most nations have chosen projected coordinate systems for use by their national mapping agencies. For many larger countries these are zonal systems like the UTM. Generally the specifications for these national coordinate grid systems include designation of a geodetic datum to be used.

The United States utilizes the State Plane Coordinate System (SPCS), in which each state is composed of one or more zones, each with its own coordinate system. Zones with a north–south orientation utilize the transverse Mercator projection, which minimizes distortion over large north–south extents (just as in UTM zones), while zones with an east-west orientation utilize a Lambert conic projection better suited for larger east–west extents (Figure 4.7). Each state zone or zone within a state has an origin located to its south and west, making all SPCS coordinates positive. The SPCS was originally based on the North American Datum of 1927 (Clarke 1866 ellipsoid) with units in feet, but now the North American Datum of 1983 (GRS1980 ellipsoid) and meter units are standard. Maps now being produced by the United States Geological Survey show NAD83 State Plane and UTM grid coordinates, with corner ticks showing the shift from the old NAD27 coordinates.

The British National Grid is an example of a single-zone national coordinate system. It utilizes a transverse Mercator projection with an origin set at 400 km north and 100 km west of 49°N 2°W, its central meridian. It is based on the Ordnance Survey of Great Britain 1936 (OSGB36) horizontal datum and the Ordnance Datum Newellyn (ODN) vertical datum. However, GPS surveying technology has increased the accuracy of the geodetic control network in Britain, and Britain's Ordnance Survey has now adopted the European Terrestrial Reference System 1989 datum as the basis for GPS mapping. As such, researchers working in Britain may find themselves dealing with three coordinate systems, based on WGS84, ETRS89, and OSGB36, but the Ordnance Survey is developing tools to ease conversions between them (OSGB, 2001).

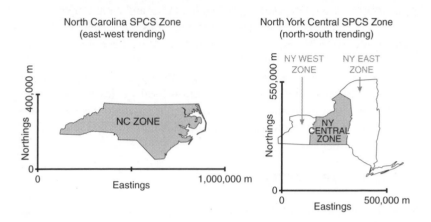

Figure 4.7. Each US state is divided into one or more zones in the State Plane Coordinate System (SPCS). Each zone has an origin to its south and east

Choice of a coordinate system to be used for a certain area can be a subjective matter. UTM is certainly an option for most of the world (all but the polar areas). The projection specifications used in many national mapping grids result in a lower level of mapping distortion than in the UTM system, so their use may be preferable in some cases. A general rule of thumb for researchers would be to utilize the coordinate system and datum employed in the other mapped data being used for a project. If the project is starting from a blank slate, adopt the coordinate system and datum used on the topographic maps produced by the national mapping agency of the country in which a research project is being undertaken.

If it is not possible to determine an appropriate projected coordinate system for use at the time fieldwork is done, working in geodetic latitude/longitude coordinates is an acceptable interim solution. Since geodetic coordinates are the fundamental coordinate system from which all planar coordinate systems are projected, utilities for performing these projections are widely available, and the collected data can be projected into a planar system at a later date once one has been selected. Many commercial vendors and other providers of spatial data distribute their data in geodetic coordinates for this reason, on the presumption that the recipients of the data will make their own decisions about what projected coordinate system to use.

Summary

At first glance, the topics covered in this chapter may seem only indirectly related to using the global positioning system in social science research. But to make use of GPS-collected information requires a basic understanding of the data itself – coordinates – and the underlying geodetic frameworks that make them meaningful. See Box 4.5 for a set of guidelines that summarizes this chapter's information about coordinate systems and datums. Careful attention paid to these few issues will eliminate an easily avoided source of problems in GPS projects.

BOX 4.5

Summary: Coordinate System Guidelines for Social Scientists

1. Geodetic coordinates (latitude/longitude on a reference ellipsoid) require specification of a horizontal geodetic datum to define an absolute location.

2. The global positioning system uses the global WGS84 datum, while most other mapped data uses other, mostly local, geodetic datums. A mismatch between datums used will result in misalignment between project spatial data from different sources.

3. Vertical positions (elevations) require specification of a vertical geodetic datum. GPS units usually report elevation as height above the ellipsoid, while other elevation data sources refer to height above the geoid (mean sea level).

4. Map projections are used to take three-dimensional geodetic coordinates and transform them to two-dimensional plane coordinate systems for ease of use on maps and in geographic information systems. The Universal Transverse Mercator coordinate system (UTM) is used worldwide, and many countries have specified their own national coordinate grids and associated geodetic datums.

5. A single, standard set of geodetic specifications including the selection of a planar coordinate system and a geodetic datum should be developed for a research project and carefully followed. Most GPS and GIS software tools allow transformation among coordinate systems, geodetic datums, and ellipsoids.

5

GPS Considerations: Getting Started

Introduction

Before beginning a project in which GPS is to be used, it is vital for the users to be aware of a number of issues associated with the equipment and the mechanics of the global positioning system itself. To the novice user, it could be overwhelming trying to decide how to begin preparing to incorporate GPS into a project. One decision that must be made is the selection of a GPS receiver to use for data collection. There are questions of price, features, and functionality that need to be asked, as well as what additional equipment is needed, if any. There is also the issue of accuracy, and what exactly that term means. Why do some units have the capability to collect more accurate data than others? Are there ways to increase the accuracy of the data? And how can understanding the answers to these questions help in the decision of which units to purchase?

These issues are addressed in this chapter. The first section is a discussion of accuracy. It deals with what is meant by accuracy, the sources of error that contribute to a decrease in accuracy, and how one can go about countering that error to increase accuracy. It is important to cover this material first, so that it will be easier to understand how the various types of receivers operate differently with respect to accuracy and error correction methods. The second section then covers the different types of GPS receivers on the market. It explains the basic features common to each receiver type that are relevant to social science data collection. The chapter wraps up with a brief discussion of data collection procedures. This is a simple overview that will be expanded upon in Part II of this book.

Understanding Accuracy and Error Correction Methods

Accuracy is the most important user element of the global positioning system, and it is the most important feature to consider when deciding which type of receiver to use. The global positioning system was developed to increase the accuracy of locating a position on the planet. This is one of the features of the global positioning system that sets it apart from older methods of determining location (see Chapter 2). Therefore, it is important to devote a detailed discussion to the level of accuracy that can be obtained by GPS receivers, the sources of error that degrade accuracy, and the common methods of correcting the error in the calculated coordinates.

GPS accuracy

It is necessary to make a distinction among the different meanings of accuracy (not to be confused with precision) when discussing the global positioning system. *Accuracy* is the amount by which a measurement or quantity differs from the actual, real-world value which it represents. However, *precision* indicates the magnitude of the difference that can be distinguished between different measurements. For example, two different tools are used to measure the length of a piece of wood. The true length of the wood is 1.2145 m. Instrument #1 returns a value of 1.2099 m, while instrument #2 returns a value of 1.21 m. Instrument #1 is more precise, but instrument #2 is more accurate, as it is closer to the true value. As they apply to GPS, we are most concerned with accuracy, given that a GPS receiver can record coordinates with almost infinite precision. That, however, does not mean that GPS coordinates are always highly accurate. The following sections on error illustrate this fact.

Both the type of receiver and the method used to collect the data can affect accuracy in different ways. Pendleton (2002) perhaps says it best by stating, ". . .a surveyor couldn't use a consumer-grade receiver to produce centimeter-level results. Similarly, a novice user couldn't produce centimeter-level accuracy with a professional-grade receiver." While it is possible to simply turn on a GPS receiver out of doors and let it automatically begin calculating coordinates, that in and of itself is not enough to ensure a high level of accuracy. There are many sources of error that can degrade the accuracy of a position, and there are ways to adjust for some of these and thereby reduce the magnitude of the error. At the same time, different receiver types have different potential accuracy levels associated with them.

Sources of error

There are six standard sources of error that contribute to the degradation of the accuracy of a calculated coordinate pair (Parkinson, 1996b). These are errors in satellite ephemeris, errors by the satellite clocks, atmospheric errors, receiver errors, multipath interference, and position dilution of precision (PDOP). The first five sources of error are cumulative, while the sixth one, PDOP, is multiplicative.

Ephemeris errors

Satellite ephemeris errors originate at the satellite and occur when the GPS signal does not transmit the correct location of the satellites. On a regular basis, the Master Control Monitor Station in Colorado updates the satellites' positions, calculates the satellites' predicted paths for the next time period, and uploads this information to the satellite constellation. These positions and calculations are stored in files called *almanacs*, which are automatically downloaded to the GPS receiver after it locates the signals from the satellites. The receiver uses the almanac data to calculate the distance to the satellite, in the same manner the ship captain calculated distances to the lighthouses in Chapter 3. Ephemeris errors are discrepancies between the predicted path and the satellite's actual path. If the satellite is in a different location from its predicted location per the almanac, the distance to the satellite is calculated incorrectly, which in turn generates inaccurate coordinates.

Satellite clock errors

Because the receiver bases its calculation on the amount of time that it takes for the signal to travel from the satellite to the receiver, any errors in the satellite's clock can affect the measurement. Each GPS satellite has four atomic clocks, which are accurate to the nanosecond. However, there is some instability in the clocks, which can cause a deviation of up to 10^{-8} secs, leading to an error of roughly 3.5 m, with an average error of 1–2 m (Parkinson, 1996b).

Atmospheric errors

There are two components of atmospheric errors. As the GPS signals leave the satellites, they first pass through the ionosphere. The ionosphere contains free electrons, which slow down the passage of the signal through this layer. In temperate zones, where the ionosphere is more stable, this delay in the signal can result in 2–5 m error. However, in polar and equatorial regions, where the ionosphere is less stable, errors can be greater (Parkinson, 1996b).

The signals must also pass through the troposphere before reaching the ground. Further delays in the signal can be encountered here due to variations in humidity, temperature and air pressure. However, these delays have much less impact on the signal, resulting in an error contribution of usually less than 1 m (Parkinson, 1996b).

Receiver errors

Rounding discrepancies between the satellite clocks and the GPS receiver clock generate a fourth source of error. In contrast to each satellite's highly accurate quadruple-redundant atomic clocks, GPS receivers contain inexpensive clocks, similar to those in common digital watches, powered by AA batteries or a similar rechargeable battery source. In fact, the "quality of the crystal oscillator is directly proportional to the quality of the receiver" (Pendleton, 2002). The atomic clocks are precise enough to determine the time up to 11 decimal places (e.g. 11:03:45.01234567890), while a receiver is less precise and may only determine time to six decimal places (e.g. 11:03:45.012346). The slight difference between the two is then a result of rounding, necessitated by the poorer precision of the receiver clock. As Pendleton states (2002), "without an accurate clock, the speed of light can be a cruel mistress." Despite that ominous proclamation, this small discrepancy often results in an error of less than 1 m.

Multipath interference

The radio waves emitted by the satellites that carry the GPS signals cannot penetrate solid objects like buildings and thick tree canopies. Instead, these objects deflect them, causing the signals to bounce around like a pinball (see Figure 5.1). This deflection is known as multipath interference, and can be a large source of error. In most cases, standing near an object like a building will lead to a small amount of multipath error, often less than 2 m. However, some large reflective surfaces such as water bodies may generate a huge amount of bounce back of the signal, leading to errors of 15 m or more (Parkinson, 1996b).

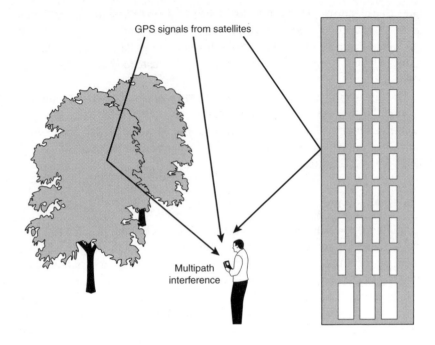

Figure 5.1. Buildings and other solid objects may deflect the satellite signals, resulting in multipath interference

Position dilution of precision (PDOP)

Position dilution of precision, or PDOP, is the sixth and most important source of error in that it is multiplicative. Position dilution of precision refers to the geometric positioning of the satellites that are being used to calculate a position. An ideal geometric positioning of satellites is one in which four or more satellites are distributed evenly throughout the sky. A poor geometry is one in which satellites are clustered together in the sky, with many satellites directly overhead or nearly so (see Figure 5.2). As the satellite geometry shifts away from the ideal situation, PDOP increases.

PDOP is a combination of two other DOPs, the Horizontal Dilution of Precision (HDOP) and the Vertical Dilution of Precision (VDOP). Although the math behind the calculations of these DOP values is beyond the scope of this book (see equation 14 in Axelrad and Brown (1996, p. 414)), it is important to note that the PDOP value has a multiplicative effect on the error contributed by the first five sources. While values of PDOP can range

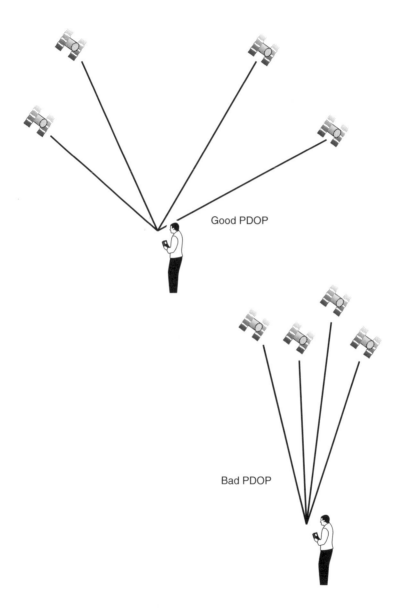

Good PDOP

Bad PDOP

Figure 5.2. Positional dilution of precision (PDOP) is a multiplicative error caused by satellite geometry

from 1 to 100, in most circumstances the values are low, with the median PDOP value worldwide being 2.7 (Parkinson, 1996a). The amount of error generated by the other five error sources is multiplied by the PDOP value to give a final overall error for any given coordinates. Therefore, if you assume 5 m of error from the first five error sources, a PDOP of 2 gives an overall error of 10 m, while a PDOP of 6 gives an overall error of 30 m. As you can see, it doesn't take much of an increase in PDOP to cause a large jump in positional error.

Overall error contributions

These errors may make it seem GPS is not as accurate as its reputation. However, when all of these error sources are combined, they still average out to around 3–15 m accuracy worldwide (Pendleton, 2002). But there are also some error correction algorithms built into all receivers that compensate for some offsets (as discussed in Chapter 3). Table 5.1, based on Table 2 in Parkinson (1996b, p. 480), contains a breakdown of the theoretical error contribution of each source.

Notice that in good conditions, the largest error is due to ionospheric effects, while the troposphere and receiver errors are minimal.

Error correction methods

Two common methods of error correction are *point averaging* and *differential correction*. Point averaging is a method that can be used with all mapping grade receivers and high-end recreational receivers (see *Types of GPS Receivers* in the following section). With point averaging, the user opens a new file, called a *waypoint*, and begins collecting data with the receiver,

Table 5.1 Error source contributions

Error source	Average contribution
Ephemeris data	2.1 m
Satellite clock	2.1 m
Ionosphere	4.0 m
Troposphere	0.7 m
Receiver error	0.5 m
Multipath	1.4 m

which is calculating new coordinates every second. During the collection, the receiver averages all of the calculated positions on the fly, constantly modifying the average location. When the user closes the file, ending the data collection, the resultant waypoint file contains one coordinate pair, which is the average position for all coordinates collected.

Differential correction is a more advanced error correction method that modifies the calculated coordinates based on the error measured by another receiver at a known location. There are two types of differential correction: *post-processing* and *real-time*. Post-processing is the original, and most common, method of differential correction. Two receivers are utilized in post-processing differential correction, both of which must be mapping grade or higher quality (see *Types of GPS Receivers* in the following section). One receiver, known as a *base station*, is located at a precisely surveyed location. The other receiver, known as a *rover unit*, is used in the field to collect data. These data are stored in files, known as *rover files*, which contain each coordinate pair that was calculated during the time that the receiver was collecting data. Thus, if a rover file is collected over the course of three minutes, and coordinates are calculated each second, then the file would contain 180 separate coordinate pairs. This is different from a point-averaged waypoint, in that the waypoint only contains the final average location rather than each calculated location.

For post-processing differential correction to work properly, the rover unit and base station must be operating concurrently and must be within 480 km (300 mi) of each other. Keep in mind, however, that the farther the rover unit is from the base station, the greater the chances that they will receive signals from different satellites. This can pose problems in developing countries, as it can be difficult to find base stations outside of major metropolitan areas. With differential correction, both receivers collect positions at the same time. The base station calculates at regular intervals, compares the calculated coordinates to the known coordinates, and computes the amount of error in the x and y directions (known as ΔX and ΔY). The coordinates, the time that they were collected, and the ΔX and ΔY values are stored in *base station files*, which are used to correct the GPS data collected by the roving receiver in the field. Given the nature of the base station files, which contain coordinates and the error offsets, post-processing differential correction can only be applied to rover files, since those files contain each coordinate pair calculated by the receiver. To correct the GPS data from the field, the user loads the rover files into a differential correction software package along with the base station files. The software then matches up each rover file point with the corresponding ΔX and ΔY computed at the base station for that same time, and modifies the rover file point accordingly. The result is a corrected rover file containing points with an increased positional accuracy of approximately 0.3–5.0 m, with the variation due to external error sources affecting the GPS at the time of the data collection, the quality of the receiver hardware and the internal algorithms it uses to calculate positions (Pendleton, 2002).

GPS users themselves can establish a base station if they have a second mapping-grade GPS receiver, and they have a precisely surveyed location they can use. The surrogate receiver can be established at a known survey location, such as a benchmark. A *benchmark* is a location usually identified with a special marker, such as a bronze disk. They typically have their latitude, longitude and elevation calculated via a horizontal and vertical control survey. If no benchmarks exist, the surrogate base receiver can be set up in base station mode and collect raw coordinate data over an extended period of time at any location in the vicinity of the study area that is relatively clear and where the receiver can be monitored and remain undisturbed. A minimum of 48 hours of measurements (continuous measurement is recommended but not required) using at least four satellites are required in order to obtain an average location that is accurate to +/- 2 m from the "true" horizontal position and +/- 4 m from the "true" elevation above the ellipsoid used (Gilbert, 1995a). The average location for the surrogate reference receiver can be used as a reference position in order to differentially correct the data collected by any roving receivers in the study area (Gilbert, 1995b). If a roving GPS receiver is used as a reference receiver, then at least one additional roving GPS receiver is required to collect raw coordinate data for target features in the study area. Alternatively, GPS users can make use of base stations established by other GPS users. Some base stations make their position files available via the World Wide Web.

All base stations collect coordinate data for base station files at a predetermined time interval. Many commercial and public base stations collect these coordinates every second daily, or during a specified period of the day. However, there are some base stations that collect coordinates at longer intervals, such as every five or ten seconds. This can cause problems when used in conjunction with rover files collected at intervals that do not precisely coincide with the station's interval, as the timestamps on the base station files and rover files may not match, resulting in few or no coordinates in the rover file being corrected. While it may be possible to overcome this problem if it arises, through the modification of certain parameters in the differential correction software package, it is important to be aware of this issue.

Real-time differential correction is a more recent development in GPS error correction. It can be used with any GPS receiver that is configured to receive the real-time signals. There are currently four common sources of real-time correction signals. The first source includes land-based beacon systems, such as the Coast Guard Maritime Differential GPS (DGPS) Service in the United States, that send out correction data via radio waves. According to the USCG website, their system provides 10 m accuracy in all established coverage areas. The beacons are strategically located to cover the entire coastline of the continental US, the Great Lakes, Puerto Rico, most of Alaska and Hawaii, and a large part of the Mississippi River Basin. Use of these signals simply requires the addition of a real-time beacon antenna to the GPS receiver. However, the range of the beacons is limited, and can only effectively be utilized within 300 mi of the stations. Other countries are beginning to develop their own coastal real-time beacon networks, modeled after the US Coast Guard system. The correction information broadcast by the USCG Maritime Differential GPS Service is free to any user who has a receiver that can pick up the signals, but it is only transmitted in the United States.

A second source consists of satellite-based subscription services. The signals are transmitted from geostationary satellites, and therefore are available worldwide. These services are expensive, but the benefit is their availability in many areas where base stations do not exist.

A third suite of real-time sources consists of free satellite-based services, such as the Wide Area Augmentation System (WAAS) and similar systems. The Wide Area Augmentation System is part of the United States Federal Aviation Administration's (FAA) GPS-based navigational system. It is a satellite-based differential correction system used to assist airplanes in navigation and landing. WAAS-compatible receivers can generate coordinates with an accuracy of 3–5 m (Garmin, 2000). Two other similar systems, the European Geostationary Navigation Overlay Service (EGNOS) and the Japanese MTSAT Satellite-based Augmentation System (MSAS), are also satellite-based differential correction systems designed to aid in aircraft navigation in Europe and Asia. All three of these systems transmit correction information from satellites over their service areas, respectively, and are free to users who have receivers capable of processing the signals.

The fourth real-time source is less common, but just as effective. This is a user transmitted correction signal. Some high-end receivers have the capability to act as base stations and calculate correction information on the fly. They can then transmit the correction data to the roving receiver by way of radio waves or cellular connections.

Types of GPS Receivers

There are multitudes of GPS receivers on the market, and new models with new applications are being developed constantly. Just like any other piece of electronic equipment, GPS receivers vary widely in price, available features, and performance. Areas of use for GPS receivers include recreation, precision agriculture, aircraft and marine navigation, precision surveying, and military. Manufacturers have made specialized receivers for each of these specialties, plus many others. It is beyond the scope of this book to describe each of these types in detail, therefore the discussion will be limited to the two most commonly used GPS receiver types in social science research: the recreational receiver and the mapping grade receiver.

Before we delve into the different types of receivers, there are some characteristics common to all GPS receivers that should be explained. The first characteristic is the channel design of the receiver. Receivers are either single-channel designs or parallel multichannel designs. Single-channel receivers can only lock on to one satellite at a time, and take a long time to fix on the minimum required four satellites. Parallel multichannel receivers can use each channel to lock on to a separate satellite signal, increasing the speed of finding those four or more satellites. For example, a 6-channel receiver can track up to 6 satellites at a time, while a 12-channel receiver can track up to 12 satellites at a time. First-generation GPS receivers are likely to be single-channel, while every receiver on the market now is a parallel multichannel, and most are also 12-channel. It is likely that over time, receivers with more than 12 channels will begin appearing in the market.

A second characteristic common to most GPS receivers is the ability to display coordinates in a variety of coordinate systems and datums. All receivers have the ability to display in geodetic coordinates (i.e. latitude and longitude) or Universal Transverse Mercator (UTM). However, the more expensive receivers are more likely to have the ability to utilize less common coordinate systems. The user can choose to display the coordinates in any supported coordinate system, independent of the processes operating within the receiver to calculate the coordinates. This allows the user to more easily compare the calculated position to coordinates on a map that uses a specific coordinate system.

A third feature is the amount of data storage, which should be considered when the data collection effort is likely to require the collection of a large number of coordinates and attribute information. Some receiver models can calculate and store a number of raw, unaveraged waypoint coordinates with user-defined IDs. Better models can perform point averaging and limited attribute recording, in addition to raw measurement and storage. Mapping grade receivers typically allow the user to do all of the above, and store rover files containing multiple coordinates and user-defined attributes for describing the target feature(s) being located (see data dictionaries below). Data storage is reported in terms of physical memory, such as kilobytes, and/or in terms of hours of data. It is important to understand how many hours of data can be stored given the project accuracy requirements and the amount of attribute data to be collected. The amount of data storage has implications for the frequency of downloading data from the receiver.

A fourth feature, *battery life*, varies from one receiver to the next. It is important to know how many hours of operation to expect from a particular receiver and battery configuration, in order to have enough spare batteries on hand during fieldwork. Keep in mind that temperature extremes will hamper battery performance as well. Most recreational and mapping grade receivers operate using two to four AA alkaline or rechargeable batteries. Some mapping grade receivers can utilize vehicle power adapters or optional 12-volt camcorder batteries, which require recharging and provide greater operation time compared to AA batteries.

A fifth feature that is not common to all GPS receivers, but is available on mapping grade receivers and higher-end recreational receivers, is multilanguage capability. This is something to consider when purchasing receivers, especially if the fieldwork is being conducted in a country that speaks a language other than English.

Recreational receivers

Recreational receivers are the most affordable GPS units on the market. They are targeted towards hikers, sportsmen, and other people that want to be able to locate and navigate to particular locations for recreational purposes. Despite the fact that they are referred to as recreational receivers, they can be quite useful for collecting data for research purposes.

Low-end recreational receivers are the least expensive of all GPS receivers, and thus they have the fewest features. The features common to these receivers are the ability to change the coordinate system and datum in which the coordinates are displayed, and a limited waypoint collection capacity. In low-end receivers, the waypoint is simply the coordinates calculated by the receiver at the time the waypoint is collected. This can be a problem, as these types of coordinates are subject to the highest amount of error, since no error correction is applied. Some low-end recreational receivers also have a multilanguage capability.

High-end recreational receivers can be relatively more expensive, but the extra cost is due to more desirable features being included in their design. High-end receivers give the user the ability to collect a more accurate waypoint through the use of point averaging. These receivers also have a data download capability. They have data ports that can be connected to a laptop or desktop computer through a specialized data transfer cable, and the files can be transferred to the computer for storage and manipulation. Some high-end receivers can perform real-time differential correction, if designed with the capability or attached to the proper antenna, and a real-time correction source is available.

Recreational receivers are easy to use and easy to find. They are designed to be very user friendly, and thus it does not take long to learn how to operate one. They are also sold on the World Wide Web and in a wide variety of stores, from outdoor recreation and sporting goods stores to department stores. Most recreational receivers come in compact sizes, making them easy to transport and store. And they are quite affordable, with low-end receivers costing around US$100, while higher-end receivers range from US$200–US$1000.

Mapping grade receivers

Mapping grade GPS receivers are developed with mapping professionals in mind. These receivers are much more expensive than recreational receivers, ranging from US$2000 to over US$10,000, but they are also much more accurate and have many more features. One of the most important features of mapping grade receivers is the capability of collecting coordinates in rover files that can be differentially corrected. Not only can point locations be collected in a rover file, but line and polygon files can be collected as well. This allows the user to travel along a linear feature, such as a road, or around the perimeter of an area, such as a lake, and collect points that delineate that feature. Those points can be differentially corrected and converted into a GIS feature data set.

As described before, a rover file contains separate coordinates for each calculated position during the time that the rover file is recording. The rate at which a position is calculated and the coordinates stored is called the *update rate*, or *logging rate*. This rate can be set by the user, and some typical rates are 1, 5, 10, or 15 sec intervals, depending on the number of desired coordinates per rover file and/or the specific application. For example, a higher log rate would be used for mapping a very sinuous road while driving the length of the road in an automobile vs. standing in one spot to record the location of a building. The logging rate also has implications for the amount of data that can be stored in the receiver's internal memory.

Mapping grade receivers give the user more benefits over recreational receivers than simply the rover file capability. They also allow the user to have greater control over the conditions under which data are collected. Such features include an elevation mask, a PDOP mask, and a signal-to-noise ratio (SNR) mask. The elevation mask is a value that corresponds to a minimum degree of altitude above the horizon, below which satellites are no longer used to calculate a position. The closer a satellite is to the horizon, the more its signal is subject to atmospheric interference. Therefore a minimum 15 ° elevation mask is recommended for most applications. The PDOP mask sets the maximum PDOP under which the receiver will calculate coordinates. For example, with a PDOP mask of 6 (a recommended setting), the receiver will only calculate coordinates when the PDOP is less than or equal to 6. If the receiver is collecting data points with a PDOP of 5.99, and the PDOP suddenly jumps to 6.01, the receiver will stop collecting data. In a similar fashion, the SNR mask sets the minimum signal-to-noise ratio for which the receiver will collect data. The signal-to-noise ratio is the proportion of the signal's information content to the level of the signal noise. If this ratio is low, much of the information in the signal is lost in the noise. At an elevation of 30 °, a signal-to-noise ratio between 12 and 20 is considered average, while an SNR of greater than 20 is very good (Trimble, 1996). A recommended SNR mask is 6, meaning that all signals with ratios below 6 will not be used to calculate coordinates. All three of these masks allow the user to set conditions that are considered poor so that the receiver does not collect data in an open rover file during those periods. This helps maintain as high an accuracy as possible in the rover data set.

A third feature common to many mapping grade receivers is the option for the user to pre-select the satellites that will be used by the receiver to calculate coordinates. This is an advanced feature, and must be used in conjunction with a software package that can read the current almanac and let the user see what satellites will be in view during the period of data collection. With enough planning, the user can select the few satellites that will provide the lowest PDOP and strongest signals for collecting the most accurate data. Once again, this is an advanced feature that is employed by experienced GPS users.

A final feature of many mapping grade receivers, and some high-end recreational receivers, is the *data dictionary*. A data dictionary is a digital form internal to the receiver that allows the user to enter attributes during

GPS field collection. The forms can be created and transferred to and from some GPS receivers or used with hand-held PCs (PDAs), and can be customized to suit particular data collection needs. They can be set up to have user-defined fields with a limited set of possible responses to choose from, a limited number of characters per field, and alerts when fields are left blank or contain invalid entries. See Chapter 9 for a more detailed discussion of the design and use of data dictionaries.

Other GPS receivers

Until recently, the other types of GPS receivers on the market were survey grade, aviation, and marine receivers. However, GPS manufacturers are constantly developing more and more specialized GPS equipment for particular industries, such as agriculture and transportation fleet management. These units are all much more specialized than the recreational and mapping grade receivers, and are much more expensive. With the exception of the survey grade receivers, these units do not convey any appreciable benefit over mapping grade receivers for social science research purposes. The survey grade receivers can achieve horizontal accuracies of less than one centimeter (1 cm), and can be useful for some research projects if the horizontal accuracy needs are high enough, but the costs are restrictive. Most projects do not require such high accuracy, making the recreational and mapping receivers sufficient.

Other GPS-related hardware

Most GPS manufacturers make accessories for their GPS units that can be purchased separately. These accessories aren't necessary for the operation of the receivers, but they provide greater flexibility in the collection and management of coordinate data.

External antennas are one of the most useful accessories that can be purchased for a GPS receiver. These antennas attach to the receiver and take over the function of receiving the satellite signals. They can be attached to long poles and extended high above the user. This allows the user to raise the antenna above nearby obstacles and gain a clearer view of the satellites.

Two other beneficial accessories for receivers are additional power supplies and data transfer cables. Power supplies can be purchased that will extend the life of the unit in the field, or that will allow it to be plugged into an AC outlet so the batteries are not drained while reviewing data indoors.

A data transfer cable will connect the GPS receiver to a laptop or desktop PC for the transfer of data files.

Selecting a Receiver and an Error Correction Technique

Choosing a receiver type and its accompanying error correction method is an important decision, and should not be taken lightly. However, it should be noted that, once the accuracy needs of the project have been established, the selection of a receiver or an error correction technique invariably leads to the selection of the other. Consider a project with low accuracy requirements (~25 m). The investigators may decide to use point averaging as the correction method. This leaves them with two choices for receivers: (1) purchase inexpensive recreational receivers to perform the point averaging, or (2) purchase expensive mapping grade receivers, which are also capable of point averaging. Most likely, they will choose the recreational receivers. However, if they operate in the other direction and first choose an inexpensive recreational receiver due to the low accuracy needs, then the point averaging correction technique is the only option.

In a different example, take a project with very high accuracy needs (2–4 m). The researchers have several choices for the receiver. They can choose a standard mapping grade receiver, which leaves post-processing differential correction as the only viable error correction technique for achieving that accuracy. They can choose a mapping grade receiver that is designed to perform real-time differential correction, and thus have also chosen the error correction method. Or they can choose a recreational receiver that is designed to perform real-time differential correction, and again have chosen the error correction method by default. To look at it the other way, by choosing differential correction, they are left with those three receiver options mentioned above.

Therefore, when selecting these two components, the decisions are interconnected. Despite that fact, and the discussions earlier in this chapter, it is sometimes easier to first select the error correction method, and then choose a receiver that will best implement the method. Some detailed guidelines are given below which should offer more insight into the decision process.

To begin, a good rule to follow is to choose an error correction method that, under the worst field conditions, results in coordinates that meet the United States National Map Accuracy Standards (NMAS), or similar standards, corresponding to the scale of the highest quality, most accurate, and finest resolution map(s) and/or spatial data layer(s) used in the research.

This rule ensures that the spatial data being collected with the GPS are compatible with the best existing spatial information and are improvements rather than degradations to the spatial database. If other geographic data have not been gathered or will not be used, and the first geographic data sets will be created from the GPS collected data, the researcher should determine a suitable baseline mapping and analysis scale and work within the map accuracy standards established for that scale.

Table 5.2 shows the United States National Map Accuracy Standards established for maps at variety of scales along with compatible GPS positioning methods, their accuracies, and relative costs. Keep in mind that these map accuracy standards apply to hardcopy maps. The conversion of hardcopy maps into digital layers, via digitization and scanning, adds additional error to the resulting digital spatial layers. The amount of that error varies with the condition of the paper map and the technique(s) used to digitize or scan the map. Designation of error correction methods, as compatible with certain groups of map scales, is based on a comparison of the horizontal accuracy for each map scale and the *upper limit* of the error correction method accuracy ranges.

Researchers should also carefully consider the *proximity* of target features when choosing an error correction method, especially when working at very large scales. Proximity refers to the nearness of features. For example, imagine an epidemiologist is interested in mapping and analyzing the spatial distribution of stagnant pools of water that serve as breeding grounds for malaria-carrying mosquitoes in a village in central Africa. Some of the pools are less than 10 m from one another. In order to capture the locations of and preserve the spatial relationships among the pools at such a large scale, a higher accuracy DGPS method is required and justified. Similarly, an earth scientist using GPS to gather ground control coordinates in order to geometrically correct a 10 m resolution SPOT satellite image, requires ground measurements that are within half an image pixel (<= 5 m) of true location; DGPS positioning methods are required and justified in this case.

Working at smaller scales opens the door to simpler, less expensive, and less accurate means of error correction. For example, if a medical geographer simply wants to map primary care health facilities at the national level in Kenya, any of the error correction methods described above can be used. However, at such small, national-level scales (e.g. 1:1,000,000–1:100,000), higher accuracy DGPS methods and their associated costs are not required or justified, even if money is no object. In other words, if a researcher is working at a scale of 1:250,000, the differences in mapping and analyses using spatial data collected to within 15 m vs. spatial data collected to within 2 m are negligible. In this situation, the researcher might use a cheaper, less

Table 5.2 Some popular map scales and the compatible positioning methods with estimated costs

Scale	Horizontal accuracy[a] (CE 90%)	Compatible positioning method (based on upper limit of positioning method accuracy)	Positioning method accuracy[a]	Estimated user costs (US$)
1:1,250	1.0 m	**Post-processing**	< 1.0 cm	15,000–20,000[b]
1:2,500	2.1 m	**DGPS** (using L1/L2 & P Code),		
1:5,000	4.3 m	**Real-time DGPS** (using L1/L2 & P Code),	1.0 cm	15,000–20,000[b]
		Post-processing DGPS (using L1 & P Code)	< 5.0 cm	4,000–11,000[b]
1:9,600	8.2 m	All of the above,		
1:10,000	8.5 m	**Post-processing**	0.3–5 m	500 upwards
1:12,500	10.6 m	**DGPS** (using L1 Code),		
1:24,000	12.2 m	**Real-time DGPS**	0.5–5 m	300 upwards
1:25,000	12.7 m	(using L1 Code)		
1:50,000	25.4 m	All of the above,		
1:63,360	32.2 m	**Average** (using L1 Code),	3.0–15 m	150 upwards
1:100,000	50.8 m			
1:250,000	126.9 m	**Raw** (using L1 Code)	3.0–15 m	80 upwards
1:1,000,000	507.9 m			

[a] *Source*: United States National Map Accuracy Standards (http://mac.usgs.gov/mac/isb/pubs/factsheets/fs17199.html)
[b] *Source*: Pendleton (2002)

complicated error correction method unless it is determined that high accuracy data are needed (e.g. focusing analyses on much smaller study sites with target features that are in very close proximity to one another) and the researcher can afford to pay for higher accuracy correction techniques. Fortunately for the users, as GPS has become more popular and the technology more advanced, the price for improved accuracy has actually dropped, with many GPS receivers under US$1000 capable of real-time differential correction. For more information on positioning methods and many other aspects of GPS, refer to Sam Wormley's global positioning system (GPS) Resources website at http://edu-observatory.org/gps/.

There are other considerations to make when purchasing GPS receivers for fieldwork. The technical specifications for the characteristics described earlier should be evaluated to determine which manufacturer and model is best suited for the project's purposes. The number of channels; the amount of data storage; battery life; the availability of multiple coordinate systems and datums, especially those used by the project; and multilanguage capability, if necessary, are five features that have already been discussed. Pendleton (2002) lists three more features to consider when evaluating the quality of a receiver. First, *antenna size* can influence the amount of local GPS signal interference encountered by the user. A larger antenna means more surface area and a larger ground plane, which improves signal reception and reduces multipath signal errors. While antennas are built into receivers, most mapping grade receivers and a few recreational receivers have external antenna connections. An optional, larger, external antenna can be purchased with a receiver, and once connected, the internal antenna is bypassed for the larger external antenna. The use of an external antenna really depends on the user application and level of accuracy. Second, GPS *receiver clocks* are important for making crucial time-distance calculations during the calculation of geographic coordinates. The quality of the crystal oscillator determines the accuracy and precision of the internal clock and thus the quality of the receiver. And third, *receiver durability* should be a consideration, especially if operation is frequent and working conditions are adverse.

Basic Steps in Collecting GPS Data

There are three simple steps to collecting coordinates using a GPS receiver, regardless of the type of receiver that you use. These steps will be discussed in greater detail in the following chapters, but this section will serve as an introduction to the methodology.

The first step is to turn on the receiver and allow it to locate satellites and acquire a signal. This process is automatic on the part of the GPS receiver, although more advanced receivers will allow the user to customize them to select particular satellites in order to optimize the signal received. As mentioned earlier, parallel multichannel receivers are superior to single-channel receivers when it comes to locating and locking in on satellites. And 12-channel receivers are the best at capturing and maintaining those signal locks in a variety of environments, from open sky to beneath tree canopies. However, in forested areas, it is important to note that the density of the canopy will affect the quality of the receiver's performance, and the receiver will likely lose its lock on the satellites if the canopy is too dense.

Once the satellites are located and the signals are acquired, the receiver will begin to calculate coordinates for its location. This allows the user to move to the second step, which is to begin collecting the coordinate data. For those receivers that collect single location waypoints, rather than performing point averaging, this involves simply hitting the appropriate button that stores the coordinates in the receiver's memory. However, for those users who are performing point averaging or collecting a rover file, a good rule of thumb is to collect a minimum of 180 positions. In most cases, this involves keeping the file open and collecting data for 3 mins, assuming that the receiver is calculating new coordinates every second.

The most flexible format for collecting data is the rover file. Rover files, which contain each individually calculated position collected while the file was open, allow for greater flexibility than point averaged files because they can be manipulated to increase accuracy (see *Error correction methods* earlier). For example, the coordinates can be plotted and outliers can be eliminated. Also, rover files can be differentially corrected, which significantly improves the accuracy of the data.

The third step is to download or record the saved coordinate information. Many receivers have the capability of downloading their files to a computer. However, some receivers do not have this capability, requiring the user to record the coordinates by hand.

Summary

Obviously, there are many things to consider before purchasing GPS receivers for a project. Understanding how the environment and equipment can alter the accuracy of collected coordinate data is just the first step. Knowing that there are ways to correct some of this error, and deciding which method best suits the user's needs, is also important. It not only allows the user to make an informed decision on which GPS receiver to purchase, but it also allows for better planning for this component of the project.

So it is necessary for all users to first determine the level of accuracy needed in the project. How to do so is discussed in following chapters. The user should also check in the study areas to determine if the facilities are available to support the GPS collection. For example, if a high level of accuracy is needed and the user decides to collect rover files and differentially correct them, a study area with no base stations within 480 km is a significant obstacle to this plan. This in turn leaves real-time differential correction as the other option, which can be quite expensive. Therefore, good planning is always the key to a successful project. Part II of this book details the steps involved in planning and executing data collection in a project.

Part II
Utilizing GPS

6
Developing a GPS Project

Because of the simplicity of many types of GPS receivers, it is very tempting for those who wish to use GPS in a social science project to assume that incorporating the technology means just handing out some receivers and telling the field team to write down the coordinates that the unit displays. As has been stressed, nothing could be farther from the truth. GPS may be a simple tool, but using it effectively requires planning and forethought before going out into the field.

The specifics of any GPS project will vary depending on the details of the project – what types of locations are being collected, whether the locations are in a jungle or a city, how many people will be in the field, and so on. However, there are certain key steps that are common to all projects using GPS. This chapter will present a brief overview of them and how they fit together. The following chapters will describe them in detail. However, it should be noted that because of the variability it is impossible to discuss everything in a way that perfectly mirrors the needs of the reader. Many of the tasks discussed in the following chapters are presented assuming a large field team in a foreign or unfamiliar location. For smaller projects that have fewer personnel, many of the tasks that are discussed as being handled by multiple people may be accomplished by one person.

Defining Data Collection Goals and Objectives

Using GPS effectively requires taking time before the fieldwork begins to plan who will collect the data, how they will be collected, and how they will be processed. Before that can be accomplished, however it is important to be clear on the goals and objectives of the fieldwork. For what objects of interest are locations needed? Why are they important? The answer to these questions will of course relate to the goals and objectives of the broader research project. And while the answers to the above questions may seem obvious to the researcher, it is an important exercise to undertake, because without a strong link between your research questions and the GPS point collection effort, there is an increased possibility that the GPS coordinates collected will not be useful.

The other reason to review the research goals and objectives is that it will help assess the project's accuracy needs. Is it necessary to know the location of the item of interest to centimeter level accuracy, or will being within 15–20 m suffice? The answer to this question has implications for what type of GPS receiver to buy, the point collection protocols and error correction techniques.

Once the accuracy requirements of the project are known, the next step is to design the data collection protocols. The work done at this stage will determine how the GPS points are recorded and by whom. If a household survey is being administered, then perhaps it is most cost effective to have the surveyor capture the GPS coordinates at the same time the survey is administered. In other cases, it may make sense to have a dedicated GPS survey team whose sole task is to collect GPS points.

Closely linked to the data collection protocol is the development of the equipment and personnel budget. At this stage, decisions will be made about what type of GPS receiver to buy, how many will need to be bought, where they will be purchased, and so on.

If this has not been answered at other stages of the planning process for the project, then questions concerning logistics will need to be answered, such as how will people be transported to and from the field, and where will they stay.

Prior to any fieldwork, the field instrument and training materials must be developed. The field instrument will be used to record the coordinates and other pertinent information. Training materials will be used to train the field team and should focus on not only the specific protocols for the project, but also provide an overview of what GPS is and how it works, as well as instruction in the use of the receivers and basic troubleshooting.

It is advisable that there be a team leader or local expert in the field who is in charge of ensuring the team follows the proper point collection protocols and can supervise the fieldwork. The team leader is also responsible for field data verification by reviewing the data collected and catching errors while the team is in the field.

At this point, the fieldwork can begin and the teams can start collecting points. Teams will collect data using the protocols presented at the training sessions, regularly checking in with the team leader to ensure that data is being collected correctly and any problems can be resolved while the team is still in the field.

After fieldwork is complete it is necessary to clean and validate the points and prepare them to be incorporated into a geographic information system (GIS) for analysis and integration with other data.

Summary

The steps involved in planning and executing a GPS project are comparable to any fieldwork and nearly all of them will apply regardless of the size or scope of the project. In many ways it is a matter of appropriate planning and being aware of the project's research goals and objectives. The following chapters will describe in detail the specific steps through planning and execution of fieldwork, and the processing and analysis of the data.

7
Project Fundamentals

Introduction

The key to any successful GPS project is preparation. Before the first person steps into the field with a GPS receiver to collect the first byte of data, it is necessary to have completed a few preliminary steps. The first step, which should always be done in any scientific study, is to define the research question or questions and outline the project goals. The second step is to determine the accuracy needs of the project. This includes deciding on an appropriate scale of analysis and then choosing accuracy strategies, equipment, and data collection procedures that best compliment that scale. Third, it is important to determine the coordinate system and datum that will be used in the study. The fourth step is to at least begin, if not complete, the development of the GIS database that will be used in conjunction with the collected GPS data.

It is essential to address these four steps before moving on to the preparations and methods discussed in the following chapters. Those chapters deal with the planning and implementation of the GPS component alone, whereas this chapter focuses on the larger picture, which is the incorporation of the spatial perspective, and more specifically GPS, into a traditional social science research project. But in addition to the four steps mentioned above, it is necessary that the researchers have a good knowledge of and familiarity with their study area. Study area familiarity is the first topic discussed, and it appears over and over again throughout the entire chapter.

Familiarizing Yourself with the Study Area

When dealing with geographic data and planning for data collection in the field, there are few things that can do more to aid or hamper the success of a project than familiarity with the study area. This is most crucial when working in foreign nations. When preparing to work in a new country, the first thing to check is whether or not that country restricts the use of GPS technology. Restrictions can range from certain strategic areas being off limits to, in extreme cases, all GPS use being prohibited. At the time of this book's publication, three countries that have varying levels of restrictions on GPS use are India, China, and Russia. While this may change in these countries in the future, and may come to pass in other countries that currently do not have restrictions, it is always best to find out for certain before planning on using the technology and spending large amounts of money in preparation of fieldwork.

Assuming that there are no restrictions on the use of GPS, there are other issues that can be addressed through a good knowledge of the study area. Security is one issue that should always be considered, whether using GPS or not. Carrying GPS equipment can make the user stand out in a crowd, and in areas that have an uncertain level of safety, this may not be a good idea. Check with the local law enforcement or your own government to determine if this should be a concern.

Logistics are a second set of issues to consider. One logistical issue is the ease, or lack thereof, of acquiring transportation in the study area and getting to where one needs to go, which often is not a trivial matter. In Southeast Asia, for instance, the monsoon season makes travel difficult, as many roads become impassable due to the large amounts of rain that fall. In other tropical areas, such as rainforest regions, rainfall is a year-round phenomenon. This means that many unpaved roads are always difficult to traverse, and may require four-wheel-drive vehicles to negotiate them. And other remote regions may only be accessible by air or river, making it costly and challenging to get from place to place. These are some examples of the environment being an impediment to travel. There are other types of logistical barriers that can be encountered, such as political barriers. In some countries, express permission is required to travel from one administrative unit to another. Knowing about these situations ahead of time is a direct benefit of close familiarity with the study area. Other logistical issues include making sure that personnel can stay in contact with each other, troubleshooting problems, etc.

A third issue is the presence of obstacles in the study that can interfere with the collection of GPS data. Is there pronounced tree cover with dense canopy closure in the areas in which the data are to be collected? Are there a lot of tall buildings? Not only can solid objects like these increase the multipath interference when collecting data, but overhead obstacles such as dense tree canopies can make it impossible to even lock on to the satellites and initialize the receiver.

All of these issues need to be taken into account when designing the data collection methodology, which will be discussed later in this chapter in the section *Establishing a data collection methodology*, and expanded upon in Chapter 8.

Defining the Research Question and Project Goals

The research question and the project goals help to structure the study and give it some direction. This is standard for any research project, and is common knowledge to all researchers who have ever written a research proposal. Despite that fact, the importance of this step cannot be emphasized enough. The purpose of this section is not to instruct users in the basics of defining research questions and project goals, but rather how to structure them in such a way as to best incorporate GPS collection and utilization of the data.

All research projects have a question that drives the research and a series of goals that must be met in order to achieve the desired results. Making the question and the goals spatial in nature is not a difficult feat (see Box 7.1), although it can be challenging for researchers who are unfamiliar with spatial data and the spatial perspective. This is merely due to the fact that most people have trouble proposing methods and analyses that they do not know exist.

When developing a project that utilizes the global positioning system, the researcher must include the spatial perspective when asking the questions and defining the goals. There are no hard and fast rules on how to do this. The best advice is to think spatially and couch all questions in a spatial context. For example, instead of asking "Does a poor diet of fast food correlate with obesity?," place it in a spatial context and change the question to "How does the proximity to fast food restaurants relate to obesity?" While the first question provides insight into what aspatial data to collect (e.g. nutrition, weight, amount of fast food eaten per week, etc.) and what analyses to perform (e.g. correlation), it doesn't take into account location. But by changing the question slightly to add a spatial component, it now defines the objectives for spatial data collection: fast food restaurants and households. So to add a spatial element when designing a research project, simply think of everything in terms of its location and how that can possibly affect the question.

Confidentiality

Confidentiality requirements can pose difficulties for collecting and using GPS data. The strength of GPS, namely providing an exact location of an

BOX 7.1

Public Health Example

A common research question among public health researchers is whether or not health care facilities are providing sufficient care to the surrounding population. In a non-spatial approach, this could be done by interviewing these facilities and gathering data on the types of services that they provide, their stocks of medicines and other medical necessities, and a list of patients over a certain period of time. However, it doesn't take much more effort to incorporate the spatial perspective into this approach. The patient data can be address geocoded to provide locations of the users (see Chapter 2). The survey teams can use GPS receivers to collect the locations of the facilities, or the providers. These data sets can be opened in a GIS and combined with other population-related data, such as census data, in order to determine if the health care facilities locations are best suited for serving their surrounding populations. Other spatial analyses can also incorporate the aspatial data gathered at the facilities. These types of analyses are discussed further in Chapter 13.

object, can also be a liability especially with highly sensitive data. In many projects that study human subjects, confidentiality is an important issue. Few people are willing to divulge sensitive information without some assurance from the researcher that the data will be protected and used responsibly. However, if GPS is used to record the exact location of these respondents, it is obviously very easy to then identify them. It is up to the researcher to balance the confidentiality needs of the project and the type of geographic data collected and distributed. While it may seem that these are mutually exclusive goals, there are some strategies that researchers can use to preserve confidentiality yet still map their data. These approaches fall into two categories: administrative control and geographic control.

Administrative control measures put safeguards into place that limit access to the geographic locations associated with the confidential data. These can run the gamut from physically storing the geographic coordinates on a different computer or in a different location, to making those who wish to use the data sign a legally binding pledge of confidentiality.

Geographic control measures rely on a modification of the coordinates to mask their true location. One approach to masking the locations is to convert the coordinates to a planar coordinate system not anchored to a particular location. This can be achieved by removing the name of the coordinate system and any zone or regional identifiers. For example, taking latitude/longitude values and converting them into UTM coordinates, but not including zone information, would preserve the relative placement of the coordinates but not allow someone to place the points on the earth. Another approach is to apply a uniform shift to the coordinates to obscure their true locations, taking care not to divulge the magnitude and direction of the shift. Since many types of spatial analyses are interested in the relative relationship between points, these approaches would facilitate analysis without compromising confidentiality. Another approach is to add a slight random error to all coordinates. Obviously this will not preserve relative position but if the project were at a large enough scale, the small amount of error introduced would be insignificant.

It is beyond the scope of this book to detail all possible approaches to preserving confidentiality with geographic data. Indeed, it is complex enough to merit a book of its own, however it is an issue that should be addressed for any project that integrates GPS with human subjects and sensitive data. Researchers are advised to consult with their institutional review board or other relevant group to determine the most appropriate steps.

Accuracy Needs

Every project has different needs in terms of the accuracy of the spatial data. The accuracy of a project is dependent on a variety of factors, including the scale of analysis, the scale of available map products, and the accuracy of the available non-GPS spatial data. In turn, the accuracy need of a project helps determine the type of GPS receiver needed for data collection, and helps define the basic data collection methodology that is the best fit.

Scales of analysis and data, and appropriate accuracy strategies

To begin, consider the scale at which the data will be analyzed. This may sound confusing at first, but is a rather simple thing to determine. Will the analysis be performed at the scale of the household or facility? The neighborhood? Census blocks or enumeration areas? Counties, districts, departments, or some larger political unit? This is the first question that must be answered, because it provides the basic indication of the accuracy need of the project. As discussed previously, projects that are working at the household level will obviously require higher accuracy than those working at the county level. For example, if the goal is to determine the distance from one house to the next, then a 100 m error in a GPS-acquired position would cause a significant problem in the analysis, since in many parts of the world houses are closer than 100 m from one another. However, if one were analyzing the distance from a health care facility to its surrounding communities, a 100 m error would not make much difference in the results.

A second factor to consider is the scale of available map products and/or digital spatial data. Regardless of whether the digital data exist or must be generated from hardcopy maps (GIS database development is discussed later in this chapter), the scale at which the products were made gives a minimum amount of error that can be expected when working with the data (see Tables 2.1 and 5.2). For example, data created at 1:10,000 scale has an inherent error of 8.47 m while data at 1:100,000 scale has an inherent error of 50.8 m (United States Geological Survey, 1947). Therefore, it doesn't make sense to collect GPS data with 2 m accuracy that will be analyzed in conjunction with 1:100,000-scale data. The amount of inherent uncertainty built into the 1:100,000-scale data will overshadow the high accuracy of the GPS data.

After considering these two issues of scale, appropriate accuracy strategies can be developed. To create an accuracy strategy, first decide the maximum acceptable error that can be allowed into the spatial data, or more specifically, into the GPS data. To continue with an example from earlier, in a study looking at distances from a health care facility to its surrounding neighborhoods, a maximum acceptable error of 50 or 100 m is adequate, given that most distances will likely be several kilometers in length. However, for a study looking at the household level, in which distances between neighboring houses is important, then the maximum acceptable error should be no more than 2 or 3 m.

Second, decide on a receiver that has the ability to provide data with that level of accuracy. And third, devise a data collection methodology that will ensure data with an acceptable level of accuracy.

Deciding on a receiver type

Now that the scale of analysis and the scales of available spatial data and data sources have been identified, and a maximum acceptable error has been chosen, the number of possible receiver types can be pared down quite a bit.

The most effective receivers on the market are those with 12 channels. For all users who are purchasing new receivers, 12-channel models are common, although some manufacturers still produce models with four channels. However, if the user has access to a pool of receivers, some of the older ones may only have 6 or 8 channels. Since having more channels means that more satellites can be continuously tracked, leading to potentially higher accuracy in the collected data, always choose more rather than fewer channels when possible.

For projects with less stringent accuracy needs, a simple low-cost recreational receiver will suffice. The least expensive receivers cannot perform point averaging, whereas a few more dollars will obtain a receiver with that capability. For those projects requiring high levels of accuracy, the best choices are mapping grade receivers that can collect rover files for post-processing differential correction, or a real-time differential correction compatible receiver, or recreational and mapping grade receivers, that can receive WAAS or other real-time correction signals. These receivers are more expensive, but can provide the required levels of accuracy. More detailed information on GPS receiver selection is discussed in Chapter 5.

Establishing a data collection methodology

Once the receivers have been purchased, how will they be used efficiently in the field to collect accurate data for the project? Careful planning is required to design an effective data collection methodology. This is just as important as devising the research question and goals, for without a well-thought-out plan for the field, all of the pre-fieldwork planning will have been in vain. This section provides a brief overview of how to create a data collection methodology. In Chapter 8, *Fieldwork Planning and Preparations: Data and Methods*, this topic is revisited in greater detail, covering GPS collection methodologies for collecting point, line, and polygon data. In the overview in this section, we only focus on point data collection.

The first thing to consider is the target for the GPS collection. What features need to have their coordinates captured? This will depend on the research question and the goals of the project. For example, in a project that is surveying households and wanting to analyze the household locations in space, the obvious target for the GPS collection is the house itself, likewise for health facilities or other public or private buildings. However, population–environment projects often utilize GPS to collect locations of different categories of land use. Therefore, large areas of varying land cover types are the targets for GPS data collection. These same projects often use GPS to develop a geodetic control network, which consists of the coordinates of known static locations, such as bridges and road intersections, which can be seen in remotely sensed imagery. In this case, the bridges and intersections are the targets of GPS data collection. And it is quite common for a project to have more than one target, resulting in multiple independent GPS data sets.

After determining the targets for data collection, the next decision to make is where precisely to collect the point in relation to the target. This is quite important, as it will make a difference in how the data is used in the analyses. When the target is a building of some sort, the ideal location for obtaining the GPS data to be collected is the center of mass of the building. In most cases, this is the center of the roof. However, it can be difficult, dangerous, and time consuming to gain access to the roof of each building in the study. So an alternative location must be chosen, one that best represents the location of the building. This can be the front door, or the center of the wall containing the front door, or the center of the north-facing wall, or on the street in front of the building, or any other location that is deemed representative of the building. But it is important that whatever the choice of location, it should be maintained throughout the data collection to the greatest possible extent.

If the target is a road intersection, the ideal spot for collecting the GPS data is the center of the intersection. Obviously, this is a dangerous proposition in all but the most remote and least trafficked of intersections. Therefore, the next best location is a corner of the intersection. Often when GPS data are collected at intersections, the coordinates will be used to georeference a remotely sensed image (see *GPS for ground truthing* in Chapter 2). In this case, it is very important that the corner at which the data are collected is recorded, along with a directional bearing to and an approximate or measured distance to the center of the intersection. This will make the georeferencing much easier.

Keep in mind that even with the most thought and preplanning possible, there will almost inevitably be situations in the field that confound data collection. Obstacles that block the signals from the satellites, such as tall buildings and dense tree canopies, can make it difficult or impossible to collect data with a standard receiver. Situations in which the satellite constellation is arrayed such that the PDOP is very high will increase the horizontal error in any calculated coordinates. If possible, it is a good idea to have some contingency plans in place to deal with any situations that can be foreseen. This relates to the discussion earlier in *Familiarizing Yourself with the Study Area*. For example, if you are collecting data at the front door of buildings and you know that there are obstacles in the study area, such as tall buildings or dense tree cover, then a contingency plan can be put in place that says in situations in which the receiver cannot locate satellites in front of the door, the user should walk 10 m away from the building and try again. If there are still problems, move another 10 m and so on until a point can be collected. Of course, it is also necessary for the user to record the distance from and direction to the building for which the point was collected.

Once the target has been identified and the location for data collection has been decided, the next step is to select the parameters under which the data will be collected. These parameters will depend on the type of receiver being used and the error correction methodology. If point averaging is being used with recreational receivers, the decision is fairly simple. It is necessary to decide on a length of time for collecting a waypoint. Our recommendation is a minimum of three minutes, or 180 seconds. This will provide sufficient data input into the waypoint to reduce the error quite a bit, while avoiding too lengthy of a time to collect each point. However, when using a mapping grade receiver to collect rover files for differential correction, there are many more parameters that can be considered. Similar to collecting a waypoint, a minimum amount of time for each rover file should be chosen. Once again, our recommendation is 180 points, which takes only three minutes if the receiver is calculating a position every second. The other parameters to select are the conditions under which the receiver will calculate coordinates and collect data for the rover, such as PDOP Mask and Elevation Mask, as discussed in Chapter 5.

In summary, it is very important to define these targets prior to beginning data collection, and have a clear idea of where the points will be collected in relation to each target. Have contingency plans in place to deal with problems arising during the data collection. And be aware that no amount of planning or foresight will account for all problems. Be flexible.

Coordinate System and Datum

The selection of a coordinate system and horizontal datum for the spatial side of the project is a rather simple task. There are two questions that, once answered, should at least reduce the choices significantly, if not leave only one choice. First, to what coordinate system and datum are the available GIS data and map products registered? All digital spatial data are registered to one coordinate system and one horizontal datum. However, most maps that have a coordinate grid have geodetic coordinates (latitude and longitude) and one or more planar coordinate systems, often UTM and/or a regional coordinate system. The same datum may be assigned to all coordinate systems, or each coordinate system may have a different datum. It is quite likely that the available maps may be in several different coordinate systems and/or datums, and it is just as likely that the digital data may be in yet another different coordinate system and datum (see Box 7.2). Only one of these many coordinate systems and datums should be selected as the base coordinate system for a project. It is recommended that the selection be a planar coordinate system, as the distance units tend to be standard units of measurement, such as feet or meters, which simplifies distance calculations. Using geodetic coordinates, with distance units of degrees/minutes/seconds, requires elliptical trigonometry for calculating distances, an unnecessary complication. Likewise, users should be aware that, if they are using a zonal coordinate system (such as UTM), their study area may fall across multiple zones, which may cause problems in effectively using the spatial data, as discussed in Chapter 4. In these cases, recording the data in latitude and longitude, and converting them to the appropriate coordinate system during post processing may be advisable.

So if the available GIS data and reference maps are registered to more than one planar coordinate system, the second question that must be answered is, which coordinate system would be most useful in the project? The recommendation here is to either use the most common coordinate system (i.e. the one shared by the most data sets of map products) or use the one to which the digital spatial data are referenced. The reason for the first recommendation should be obvious. Using the most common reference system minimizes the number of coordinate conversions that have to be done. The reason for the second recommendation is that it is sometimes easier to perform coordinate conversions when converting map products to digital data sets than it is to transform existing digital data from one reference system to another.

The final issue to keep in mind is that, by default, most GPS receivers will report coordinates in latitude and longitude, using the WGS84 horizontal datum. Some receivers have the ability to display and export coordinates in a variety of different coordinate systems and datums, which is useful when using the receiver to navigate with hardcopy maps or linking to data layers in a GIS. Most software packages that are designed to interact with their receivers and download the data – for those receivers with this capability – have the capability to export the GPS data in many different coordinate systems. If this is not an option with the software being used, there are third-party coordinate calculator software packages that can be used to do the conversions.

BOX 7.2

Multiple Coordinate Systems and Horizontal Datums

It is possible to have a variety of coordinate systems and horizontal datums among various available map products and digital spatial data products. This is often due to a change in cartographic standards over time, with the difference showing up in products from different eras. For example, many older maps of North Carolina are in the US State Plane Coordinate System, North Carolina Zone, with NAD27 horizontal datum, and the distance units are feet. Newer North Carolina maps are also in the US State Plane Coordinate System, North Carolina Zone, but the horizontal datum is NAD83 and the units are meters. Still other maps, made by organizations not affiliated with the state government, may produce maps in the UTM coordinate system with the WGS84 horizontal datum and units of meters.

Developing a GIS Database

Projects that incorporate the spatial perspective need to have a base of spatial data. In most cases, a GIS database is the most appropriate way to structure the spatial and aspatial data that will be used in conjunction with the GPS data. A GIS is a set of tools that analyze a variety of data, both spatial and aspatial. The spatial data can include, but are not limited to, vector and raster GIS layers, remotely sensed imagery, and GPS data. The aspatial data can include, but are not limited to, descriptive attribute data associated with the spatial data, survey data acquired through interviews or survey forms, and other ancillary data that provide insight into the study area. The strength of a GIS is that all of these different types of data can be placed within one software package, integrated together, and analyzed as a group.

The development of a GIS database for a social science research project can be a considerable task, depending on the location of the study site and the goals of the project. In countries with well-established digital data repositories, the principal job of the database developer is contacting people and organizations with the data and acquiring it. However, in countries with little or no existing digital data, the difficulty of building the database increases significantly. This usually entails building the GIS data sets from the ground up, using techniques such as digitizing of hard copy maps and air photo interpretation. Regardless of the manner in which the GIS database is constructed, it is important to have a large portion of it complete before beginning fieldwork. It is also just as important to understand the limitations of the data that are available and/or will be created. Being familiar with the scale and accuracy of the data, as discussed above, will inform the decisions about accuracy needs, GPS receiver types, and collection methodology.

As mentioned in the previous section, it is quite likely that data coming from a variety of sources will not all be in the same coordinate system and datum. However, it is necessary to have all data sets in the same coordinate system and datum prior to beginning analysis. So how is this accomplished? There are a variety of software packages that can transform spatial data from one coordinate system and datum to another. Most robust GIS software packages, such as ArcInfo, ArcView, and MapInfo are fully capable of performing these transformations. There are also other software packages, such as Blue Marble's Geographic Calculator, which can convert spatial point data and individual coordinates from one system and datum to another. By using one of these software packages, existing spatial data can be transformed to the project's chosen coordinate system and datum.

But it gets a little trickier when creating data sets from hardcopy sources, such as digitizing maps. For example, if the map is referenced to the Philippine Reference System 1992 and the Luzon datum, and your project is using the Universal Transverse Mercator (UTM) system and the WGS84 datum, how do you get the resultant digitized data set into the project coordinate system and datum? A possible solution is to digitize the map features in the Philippine Reference System 1992 and the Luzon datum and then transform the data set in a GIS software package.

When developing a GIS database, its contents and the complexity of its structure are dependent on the needs of the project. A project that only calls for the use of point, line, and polygon data will result in a much simpler database with fewer data sets than a project that utilizes remotely sensed imagery as well.

Consider, for example, a sociological project that is looking at relationships between the placement of bus stops and the socioeconomic status of the neighborhoods in which they are located. At the simplest level, this analysis would only require two data sets: bus stop locations, possibly collected with GPS receivers, and census block data. Slightly more complex analyses along these lines might necessitate the addition of a roads layer. But all in all, this is a rather simple database.

At the other extreme, consider a population–environment project in a developing country that is trying to understand patterns of land use change in relation to activities by major corporations. This database would require many more data sets than the previous example, with a much more complex inter-action between the different elements. It may contain a combination of points, lines, polygons, remotely sensed data, GPS data, and aspatial data. All of these are typical components of a GIS database. The remotely sensed data would be used to analyze the land use change over time. The point, line, and polygon data would provide some contextual information such as roads, po-litical boundaries, populated places, hydrologic features, and areas owned by the corporations. The GPS data could be locations of houses at which surveys were conducted to find out residents' concerns over corporate activities, or they could be locations of the actual activities themselves. And the aspatial data would be the attribution information that is to be analyzed in conjunction with the spatial data, such as the results of the surveys or descriptions of the locations of corporate activities. Obviously, with such a complex and varied set of data, the analyses are able to be more complex as well. But the time and effort involved in setting up a database of this magnitude is much greater than that required to develop the database described in the previous example.

Regardless of the database structure and its contents, the development of a GIS database is a task that should not be overlooked. Simply collecting GPS data to locate some phenomena of interest is not enough to perform useful spatial analyses. The GPS data must be accompanied by other spatial data sets in order to give the locations of the phenomena some geographic context in which they can be analyzed. The accuracy and validity of spatial analytical results that come out of any social science research project are highly dependent on the quality of the data in the GIS database. Proper foresight, planning, and care must be taken when developing these databases to ensure good results.

Summary

It should now be apparent that the incorporation of GPS and the spatial perspective into social science research is not something to take lightly. It requires as much thought and pre-planning as any other part of the project, and also requires a good working knowledge of spatial data, geographic information systems, and the global positioning system.

This chapter has introduced the fundamental issues necessary to begin implementing a spatial aspect in a research project. The following five chapters discuss in greater detail how to put all of this to work, from pre-fieldwork planning, through the actual fieldwork itself, and finally wrap-ping up with the post-fieldwork processing of the data.

8

Fieldwork Planning and Preparations: Data and Methods

Why Are Fieldwork Planning and Preparations Important?

Proper planning and preparations for fieldwork are essential to the success of the data collection effort, and the best measure of success is in the quality of the data. But all the planning in the world cannot guarantee trouble-free data collection and error-free data. In other words, Murphy's Law always applies: if something can go wrong, it will. Invariably, methodological problems, technical glitches, equipment failures, logistical snags, personnel issues, weather events, political situations or other mishaps and setbacks arise, many being beyond the control of the researcher. This is true of any fieldwork.

Fieldwork is a major investment in time and resources, and it is during this time that data quality is most susceptible to preventable errors. While there are no guarantees in fieldwork, sufficient planning, field preparations, and foresight on the part of the researcher regarding decisions one must make before reaching the field, as well as the types of issues one might encounter in the field, will help insure a productive fieldwork session with high quality data, efficient use of available resources, money and time, and minimal surprises. Thus, the importance of sound fieldwork planning and preparations cannot be stressed enough, and this applies to projects of all sizes and scopes.

Fieldwork planning and preparations are multifaceted processes that can easily take more thought, time, and effort than the actual fieldwork. In fact, the subject probably warrants an entire book to cover the topic in sufficient detail. We have attempted to condense the topic into three related chapters corresponding to our organization of fieldwork planning and preparations, beginning with data and methods in this chapter.

Data

Spatial data characterization

Fieldwork planning begins by considering the research questions and determining the set of real features and phenomena involved, the spatial relationships among those features and phenomena, and the mapping techniques and spatial analyses one might use in order to answer the research questions. Three basic questions should follow concerning the representation of real landscape features and phenomena as spatial, or geographic, data within a GIS. They are:

1   what is to be located?
2   how is it to be described?
3   how is it to be symbolized?

By answering these questions, the researcher begins to characterize the spatial data that is to be collected, laying the foundation for spatial database design and the development of the data collection methodology.

What is to be located?

While this question was addressed earlier as part of the project fundamentals in Chapter 7, it is reiterated here to underscore the importance of defining exactly what needs to be located during the fieldwork. The researcher should gather all existing hardcopy maps, digital spatial data, and other available data, consider these sources along with the research questions, and develop a list of features and phenomena to include in the spatial data collection. These are the real landscape features and phenomena that the researcher wishes to capture and input into the spatial database in order to answer the research questions via mapping and/or spatial analysis. For example, a researcher interested in community mapping might use GPS receivers to locate houses, wells, roads, rivers, streams, canals, village centers, agricultural fields, and important community boundaries for a number of communities in a region. By clearly defining the features of interest, the researcher can determine if GPS methods are required to capture those features or if they can be adequately located and converted to digital form using other sources and means, such as the digitization of topographic maps. Topographic maps might provide the locations of rivers and streams in sufficient detail and completeness for the research, although recent or frequently changing human-made features, such as houses, wells, canals, and roads, might need to be located using GPS receivers.

The researcher should also understand how changes in the research questions affect the type and number of features that must be located and mapped and the additional effort it will entail. Imagine a researcher interested in determining the straight-line distance between towns and nearby health clinics in a study region. In this example, a geographic location and a *key*, or *unique, identifier* for each health clinic are required. The result in the GIS is a geographic data layer consisting of one type of target feature, health clinics, and one descriptor, the unique identifier. If the researcher is interested in mapping the distribution of health clinics relative to those same town locations, at a minimum, the locations of the health clinics and towns as well as a key identifier for each are needed. If the distance along the roads from each of the health clinics to each of the towns is desired, the road network must be mapped in addition to the locations of, and key identifiers for, the health clinics and towns. These examples are relatively straightforward, but they do demonstrate how the collection of spatial data can quickly become more complex and resource consuming with seemingly innocuous modifications to the research questions.

How is it to be described?

Features and phenomena displayed on a map, or located using GPS technology, are described first and foremost by their geographic position. To derive any meaning from a set of feature locations via mapping, pattern analysis, or some other spatial analysis, further description of each geographic feature is required, usually the result of recorded observations, responses, or measurements made at the feature locations. The information that is linked to a feature location and describes a geographic feature's properties and behavior is known as *attribute data*. The type of information collected as part of the attribute data is mainly dependent upon the research questions. At a minimum, and regardless of the research questions, a unique, key identifier needs to be explicitly linked to each unique geographic feature and phenomena. The key identifier can be a numeric code, a name, or some combination of the two, and it serves three primary purposes in the database:

● provides a basic, unique identification of the geographic feature;
● allows geographic features to be queried once they are input into the spatial database;
● allows additional attribute data from field data collection and other tabular sources to be linked to the geographic features, provided that matching key identifiers are present in the records of the ancillary data.

Building upon the previous clinic example, health clinic and town names and/or numeric codes might be used as key identifiers for the health clinic and town features. Road segments might have unique numeric codes and/ or street names and address ranges associated with them. Additional attributes should be collected for, or linked from other available sources to, each geographic feature as the research questions dictate. This opens the door to more informative mapping and more sophisticated spatial analyses down the road. For example, mapping and examining the spatial pattern of health care services at each clinic requires attribute data on each clinic's services. Estimating travel times along roads between health clinics and towns requires road attribute data (e.g. surface type, speed limit, number of lanes, traffic volume) and traveler inputs (e.g. mode and speed of travel). This is in addition to health clinic and town identification. As these examples reveal, feature locations are not very useful without attribute data, or feature description. However, meaningful mapping and spatial analyses within a GIS are dependent not only on feature locations and their attributes, but also on the visual renderings of the features in the GIS.

How is it to be symbolized?

Real landscape features and phenomena are represented graphically in a GIS as spatial objects that take the shape of geometric points, lines, polygons (areas), and three-dimensional forms (surfaces and objects that have volume) that are geographically referenced to known coordinates. When developing the spatial database and the data collection protocols, one must consider how the features might be appropriately represented. The symbolization of a feature or phenomena in the GIS is closely tied to:

- the geographic variation in real feature and phenomena shapes;
- the ability of the researcher to capture those shapes;
- and the nature and scales of analysis.

Real features and phenomena have dimensions that define their shape. These shapes vary across the landscape and, depending on the feature, may change over time. To the extent possible, features and phenomena should be symbolized within the GIS much like they exist in the real world, although a perfect digital representation of real features and phenomenon has yet to be created. Therefore, a model that most closely resembles the actual geographic variation in feature form and function is preferable. For example, a researcher is interested in examining the spatial relationships among well and stream gauge data, and the streams, ponds, forest stands and their characteristics, within a small watershed. Well and stream gauge locations might be appropriately represented as geometric points within the spatial database; it would not make much sense to represent the other higher-dimension features as points. The streams would be represented as lines, whereas ponds, forest stands and the watershed would be represented as polygons. The watershed might also be represented as a three-dimensional surface, provided sufficient elevation data (z-values) are available for a number of geographic locations.

The ability of the researcher to capture the spatial data necessary to represent features and phenomena should also be considered. Building on the watershed example, limited resources and/or inaccessibility may necessitate concessions on the part of the researcher as to which features are located and/or how they are located. As a result, locating well and stream gauges using GPS may take priority over delineating streams, ponds and forest stands with GPS, as the latter features may be captured using aerial photographs and/or topographic maps. However, aerial photographs and topographic maps may be unavailable for the study area, or they may be available but not be as current as the GPS data, while streams, ponds, and forest stands may have changed or been altered since the publication of the photographs and maps. Furthermore, given a large number of ponds within the watershed, the researcher might decide that instead of capturing the perimeter of the ponds in order to create a polygon spatial object, there may only be time to record a single location at the edge of each pond with an estimate of distance and direction to the pond center.

In other instances, the symbolization of features and phenomena as points, lines, polygons, or three-dimensional shapes may have less to do with resource constraints, as above, and more to do with the nature and scales of analyses. For example, detailed hydrologic modeling of a watershed may dictate that the perimeter of every pond and forest stand be captured in order to create polygon objects in the spatial database. In addition, pond depth samples and tree canopy height measurements might be desired for the generation of three-dimensional pond and forest stand representations that fit into a three-dimensional elevation model of the watershed. If the scale of analysis and the data collection methods are sufficiently detailed, streams, rather than being symbolized as single lines, might be represented as double lines (left and right banks), linear polygons, or as three-dimensional models (additional stream depth sampling required).

Methods

Data collection methodology

The next step in fieldwork planning is to address the question of how features and phenomena are to be located. (For simplicity, the features and phenomena to be located are hereafter referred to as target features.) Put simply, one or more coordinates must be recorded at, or in the vicinity of, each target feature along with spatially linked attribute data. The absolute accuracy of the coordinates is dependent upon the error correction technique (i.e. point averaging versus differential correction) used to derive the coordinates. The number of coordinates required is based on the shape of the target feature, the desired spatial object representation for that feature, and the spatial data capture method used. The type of attribute data collected for each feature is based upon the research questions, and the people participating in the actual data collection depend on the nature and scope of the research. The data collection methodology should therefore consist of the following:

1   a GPS error correction method that satisfies the accuracy requirements established for the research project;
2   well-defined data capture protocols that are repeatable and correctly and consistently applied throughout the fieldwork;
3   procedures for ensuring data quality.

As the choice of error correction has implications for the data capture protocols, it should be one of the first decisions made in developing the data collection methodology. Data capture protocols follow with procedures and guidelines for collecting spatial and attribute data for a variety of target features. Examples are provided for the reader to consider and apply to his/her own research needs. Data quality is enhanced via a number of procedures implemented during the data collection and data entry phases of a project. These procedures are addressed thoroughly in Chapter 10.

Selecting an error correction method

In a nutshell, higher accuracy spatial data are better, but they are not always justified, required, or affordable. As stated previously, researchers must collectively consider the application of the GPS, the accuracy needs of the project, and the budget for GPS equipment, data processing, and differential correction services. It is vitally important to specify the GPS data error correction method in the project and fieldwork planning stages. This will ensure that all necessary steps to implement the desired correction method will be handled properly. For example, if raw or point averaged coordinates are to be collected, some method of recording the coordinates must be chosen. If real-time differential correction is to be used, a subscription service must be identified and purchased or a free alternative service used, if available. If post-processing differential correction is to be used, an existing base station must be located or a new base station must be established by the project, and the base station files must be obtained. Refer to Chapter 5 for a complete discussion of GPS error correction methods and guidelines for selecting an appropriate method.

Data capture protocols

Data capture protocols are the second component of the data collection methodology. Point collection protocols were briefly discussed in Chapter 7. Here, we expand on point collection and discuss line, polygon, and three-dimensional collection as well. Protocols describe how to correctly and consistently locate and describe target features in the field. As such, data capture protocols should include guidelines, decision rules, and contingency plans for capturing spatial data and recording feature attributes for the target features specific to the research. Where should one stand in relation to a target feature when recording its location? If a target feature is inaccessible by foot, but in view, how can one record a geographic position for that feature? How can one record the perimeter of a lake or some other feature without walking around the edge and recording coordinates with a GPS receiver? Is a separate GPS team required to collect spatial data or can spatial data collection be easily incorporated into a nonspatial survey? What considerations should be made when recording spatially linked attribute data? Protocols are established in order to answer these types of questions and address such issues.

Before addressing the protocols for capturing the location and attributes of point, line, polygon, and three-dimensional features, there are some data capture fundamentals common to all projects using GPS as a GIS data capture tool that warrant some discussion and should be considered in fieldwork planning. The first subsection below discusses these common issues. Spatial data capture protocols for various feature types are addressed in the second subsection followed by feature attribute considerations in subsection three.

Data capture fundamentals for planning

Geographic reference

Chapter 7 covered the selection of an appropriate coordinate system, datum and units for the project and the GPS data collection. It is important to keep this in mind throughout all phases of planning, fieldwork, post-fieldwork processing, and analysis.

A clear view of the sky

One of the basic requirements when using GPS to calculate geographic co-ordinates is that the receiver must have a clear view of the sky. The receiver uses its antenna to receive signals from orbiting GPS satellites. Overhead obstacles such as dense tree canopy cover, tall buildings, steep canyon walls, mountains, and even the GPS operator's body (for example, a hand covering the antenna) can interfere with or completely block the signals from one or more satellites, hampering the performance and accuracy of the GPS receiver in some cases, or rendering the GPS receiver useless in the worst cases. Projects working in environments where complete blockage of signals might occur may benefit more from traditional surveying techniques. In situations where the obstacles are not difficult or impossible to overcome, an external antenna can be very useful when mounted on a device that will allow the receiver to be raised above the obstacles.

Rover file/waypoint naming conventions

Most GPS receivers allow the user to define the names of waypoints and rover files stored in the GPS receiver memory. Researchers should develop logical naming conventions for waypoints and rover files and instruct the data collection teams to adhere to these conventions. While most GPS receivers allow the user to manually assign a file name, many will automatically assign a name based on a sequential number or time/date. If multiple teams will be gathering GPS coordinates during the same time period, it is important to ensure that file names will be unique between teams. Otherwise, GPS data files collected by different teams may have identical file names, which can be problematic when the files are downloaded to a computer and organized in the database.

GPS memory and data downloading

The internal memory of GPS receivers is limited and varies among receivers. Consult the GPS receiver's manual and technical specifications for these limits. If planning to store spatial and attribute data in the GPS receiver's internal memory, be sure to develop a plan for downloading the data periodically to a computer. After the data is downloaded safely, the receiver's memory can be cleared for more data collection. Data should be downloaded on a daily basis if possible, though some projects may require more frequent downloads as a result of the intensity of data collection and the techniques used to gather the data. In instances where a data collection team fills up the receiver memory very quickly, the team might carry its own laptop or other data logging device for downloading data in the field, or utilize a second receiver once the first receiver's memory is full. See Chapter 10 for more information regarding data download and backup.

Who will collect the spatial and attribute data?
Data collection using GPS technology may be a primary or a complementary component of a data collection effort, depending on the focus of a project. Mapping projects may be concerned with locating features and recording a few attributes about those features, thus the spatial information and mapping are the focus. Other projects may incorporate GPS receivers into new surveys or existing longitudinal surveys, using GPS to fit the study sites with spatial coordinates while at the same time gathering other types of data about each site via one or more detailed questionnaires or data forms. The GPS collected data are not the focus; rather, the spatial data are complementary to the other data being collected. In planning a data collection effort using GPS, it is important to determine whether GPS data collection is done as a separate spatial survey or is folded into a primarily non-spatial survey. This determination has implications for the amount and type of personnel, the data forms and training needed, and the amount of time and money invested in the fieldwork. For example, by training a demographic survey team to gather GPS data for the communities visited, the need for a separate spatial data collection team is eliminated. In some instances a separate team or person, specifically trained and charged with collecting GPS data, may be warranted, as in when locating previously surveyed communities in order to retrofit them with spatial coordinates.

Spatial Data Capture

Target feature geometry can be simple or highly complex. Representing a target feature in a spatial database necessitates a generalization of its real world shape into one of five spatial object types: point, line, polygon, three-dimensional surface, or three-dimensional object with volume. The basic building block for these spatial objects is the point, represented as a single coordinate pair. Higher dimension objects are created using two or more coordinates and the addition of altitude, elevation, or measurement z for surfaces and objects with volume. Figure 8.1 illustrates how higher dimension objects are constructed from points. Two or more points can be linked together to form a one-dimensional line (Figure 8.1a); the points become the vertices between the line segments. Three or more points can be linked together to form a closed, two-dimensional polygon that has area (Figure 8.1b); the points become the vertices at the corners of the polygon. Points can also take on a z-value and be used to create three-dimensional surfaces (Figure 8.1c) and three-dimensional objects with volume (Figure 8.1d). The GPS receiver is used to locate a point, or a set of points, in the geometry of, or in the vicinity of, the target feature.

When developing data capture protocols using GPS receivers, the accessibility and shape of the target features are primary considerations. Initially, the researcher should think about all target features in the study area and the possible obstacles to reaching them to record their geographic location. In other words, one should consider the ultimate position of the data collector and GPS receiver relative to the target feature's location and the target feature's geometry. Decision rules and contingency plans should be developed as part of the protocols to deal with accessibility issues and other obstacles to spatial data capture. By establishing systematic rules for locating target features, the data and methods remain consistent and repeatable throughout the survey. If target feature accessibility becomes an issue, other mapping tools and distance measuring devices can be used in conjunction with the GPS receiver to obtain coordinates for those target features, as discussed in the following section.

Other considerations when developing spatial data capture protocols include the proximity of the target features to one another, the number of coordinates to calculate and record in each file for point locations, and the use of sketch maps. The two former considerations relate more to the type of GPS receiver and the error correction method used in the data collection methodology. They are discussed in detail in Chapter 5. As for the latter, a useful mapping and data capture tool that may be included as part of the spatial data protocols is a sketch map. *Sketch maps* are created and used to record both spatial and attribute information about target features. A sketch map of a target feature and its surroundings provides spatial context for documenting the position of a GPS measurement relative to the target feature's location and geometry. Features can be labeled and other attribute information can also be recorded on the sketch map. For example, a sketch map might be drawn showing the boundaries of a piece of property, an adjacent road, and the locations of two GPS measurements taken at the two corners of the property that are adjacent to the road. The road might be labeled with its name and the loca-

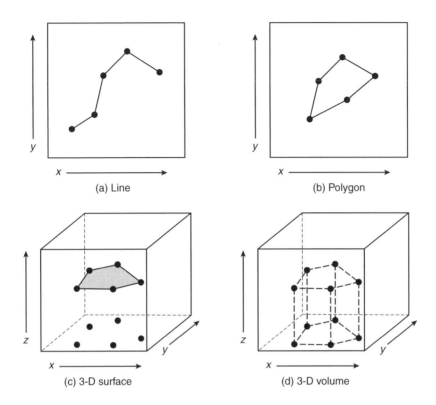

Figure 8.1. Higher dimension spatial objects created from point locations

tions of the GPS measurements should be labeled with their rover file name or waypoint name. Sketch maps should be drawn with some sort of reference direction indicated, such as a north arrow, and they are typically not drawn to scale. Fortunately, sketch maps can be used with any of the target feature types discussed in the following sections, and they are particularly useful for diagramming complex features.

The following four subsections, organized by type of spatial feature, provide the reader with typical protocols for capturing a target feature's location and shape. Examples are provided in each subsection to illustrate proper data capture protocols for a variety of field situations. Practical instructions for operating a GPS receiver are not presented here. GPS practicalities are presented in Chapter 11, and users should consult their GPS receiver's manual for specific instructions for calculating coordinates and recording GPS coordinate data to files.

Point features

Point features are target features that are represented in the spatial database by a single geographic coordinate pair. These are the simplest types of target features to locate in terms of spatial data capture. Some examples include houses, health clinics, community centers, village locations, stream gauges, wells, and weather and air quality monitoring stations. Assuming that the target feature is accessible, the researcher needs to decide literally where to position the GPS receiver relative to the target feature's location and geometry. For example, a water quality researcher, leading a project that conducts household surveys and gathers well water samples in several rural towns across a region, is interested in determining absolute locations for the towns, the respondents' houses, and their water wells. The resulting geographic layers will be input into a GIS for spatial analysis.

For town locations, the researcher may instruct the data collector to stand with the GPS receiver at the northernmost corner of the main intersection of each town, or alternatively, on the sidewalk directly in front of a significant landmark, such as town hall. An ideal position for the data collector to record the absolute location of a well would be in the center of the well structure, either standing on the well cap, placing the GPS receiver on the well cap, or holding the GPS receiver over the well. For obvious reasons, these approaches are not practical when recording the location of a house. One option is to record the location approximately five meters away from the front door in a direction that is perpendicular to the front door. If this position is obstructed, the data collector could move approximately five meters to the right as he/she is facing the house. Alternative approaches might have the data collector always positioned in the approximate center of the front lawn or at the entrance to the driveway. Technically, the coordinates obtained for the well and the house are absolute locations, however the coordinates obtained for the house are not the true location of the house since the coordinates were recorded in a position that is relative to the house location and its geometry. As for the wells, the data collector is able to position the GPS receiver in direct contact or just above the well, providing an absolute location that is also a true location (subject to the stated error of the GPS receiver and error correction method used).

If the target feature is inaccessible but still in view of the data collector, the data collector may record a coordinate pair in the vicinity of the target feature and augment those coordinates with a direction, distance and, in some cases, slope angle to the desired data capture position. Using the previous example, the data collector might be instructed to determine the direction and distance from the GPS receiver position to a point that is well within the walls of the house. Direction can be measured using a *compass* and distance measured using *meter tape*. The resulting offset measurement can then be applied to the GPS-derived coordinates, resulting in augmented coordinates for the house location.

A more advanced example using GPS receivers and other mapping tools to augment GPS coordinates is a land cover classification project mapping swidden (i.e. shifting cultivation, slash and burn) agricultural fields to gather ground truth data for use with satellite imagery. Within the study area, numerous swidden fields are scattered on the mountainsides above a limited number of roads that traverse the valley bottoms. Reaching many of the fields on foot is not feasible given the combination of rugged terrain, dense forest, and limited time in the field. How can the researcher quickly and efficiently map a location for each of the fields with some degree of accuracy without standing in each field with the GPS receiver? Using the same methods in the previous example but with a few more measurements and calculations, it is possible.

For this application, the GPS user must be able to see the swidden field

---

BOX 8.1

What is a Compass?

A compass is a mapping and navigation tool that measures angles in the horizontal plane. The needle on a compass always points to magnetic north, which is often used as the reference angle at 0 ° on the compass dial. Note that the location of magnetic north varies over time and the direction of magnetic north varies depending on the compass user's location. Magnetic north is not the same as the true north specified on many reference maps. Depending on where one is in the world and the reference map that is being used, most compasses need calibration to adjust for the difference, or *declination*, between true north and magnetic north; subsequent readings are then relative to true north. Compasses are useful for orienting oneself in relation to magnetic and true north and for determining a general direction or a specific azimuth (0–360 °) or bearing (0–90 ° within a compass quadrant) from one location or object to another. Some GPS receivers have a digital or magnetic compass as one of their features.

BOX 8.2

What is Meter Tape?

Meter tape is an instrument used to accurately measure short distances. Metric tapes are sold in reels of measuring tape in lengths ranging from 15 to 100 meters. Tape is typically marked off in meters, decimeters, and centimeters. Distances longer than the total length of the tape can be measured by marking the position at the end of the tape, making a subsequent measurement from the marked position (repeat if necessary), and then adding the results from all measurements. Care should be taken that the tape is pulled taut between measurement points and that distances are accurately read from the tape.

---

clearly from his/her vantage point, presumably somewhere along the road. Four measurements are made using the GPS receiver and other mapping tools. The GPS receiver is used to calculate the geographic coordinates where the user is standing. A compass is then utilized to determine the azimuth (0–360 °), or "look direction," to the field. Since the user is located in the valley bottom and below the field, a distance from the user to the field must be measured along a slope angle, or "look angle". You may also see slope angle referred to as the angle of inclination. The look angle between the imaginary horizontal plane that the user is standing on and the field on the mountainside is measured using a *clinometer*. A *rangefinder* is then used to measure the slope distance, or "look angle distance," from the user to the approximate center of the field. These four measurements–GPS coordinates at user position, azimuth, slope angle, and slope distance–are input into five simple equations (Table 8.1) in order to calculate an augmented GPS coordinate pair (x,y) for the approximate center of the swidden field.

Figure 8.2 illustrates the important difference between the slope distance and the planimetric distance, or true horizontal distance, between the GPS user and the swidden field. Without accounting for the slope angle in calculating augmented coordinates, the slope distance between the GPS user and the swidden field, measured along the slope angle, is treated incorrectly as the straight-line distance along the horizontal plane. The resulting augmented coordinates place the swidden field further from the user than the field's true horizontal distance and actual location, which could be a disheartening surprise once the data are mapped with the satellite imagery and other GIS data layers.

BOX 8.3

What is a Clinometer?

A clinometer is an instrument that measures angles from horizontal to a target in the vertical plane, the angle of inclination. Measurements can be made in terms of degrees (0–90; horizontal to straight up or straight down) and percent slope (0–100). Some higher quality compasses include a clinometer as one of its features. Clinometers are not as accurate as professional surveying instruments, however they are relatively inexpensive and provide sufficient estimates for most social science mapping purposes and general navigation. Clinometers can also be used in conjunction with mathematical/geometric formulas in order to estimate the height of objects. Please consult the compass or clinometer user manual for detailed instructions.

BOX 8.4

What is a Rangefinder?

A rangefinder is an instrument that measures the distance from one location or object to another. Optical and laser rangefinders are available. Optical rangefinders measure the distance to targets using the focal angle between two lenses mounted in the rangefinder. Optical rangefinders are most accurate at measuring horizontal distances up to approximately 200 m with an accuracy that is +/- 3 m depending on the distance of the target.

Laser rangefinders incorporate microwave or infrared lasers for measuring distances to targets. Generally, laser rangefinder accuracy is around +/- 1 to 2 m for ranges from 400 m to 1.2 km. Under ideal conditions, measuring horizontally to a highly reflective target, certain models claim to have a maximum range of 1.5 km with +/- 1 m accuracy. More advanced laser rangefinders have a clinometer and digital compass built in and boast 1 mm accuracy under certain conditions. When considering a laser rangefinder, be sure to note the stated range and accuracy for measuring to nonreflective targets and under different field conditions. For example, measuring in rain and through brush can impair laser rangefinder performance. Some rangefinders also have magnification capabilities similar to binoculars.

Table 8.1 Equations for deriving augmented geographic coordinates

1. Horizontal distance = (*Slope distance*) × cos (*Slope angle*)

2. Change in Easting (Δx) = (*Horizontal distance*) × sin (*Azimuth*)

3. Change in Northing (Δy) = (*Horizontal distance*) × cos (*Azimuth*)

4. Augmented Easting = (*GPS Easting*) + (Δx)

5. Augmented Northing = (*GPS Northing*) + (Δy)

There are several added bonuses to this type of method. The measurements and equations discussed in the previous example can be applied to map remote targets located *below* the user with no change in methods other than the user looking down instead of up when making measurements. Also, many targets can be mapped from a single user position provided there is a clear line of sight between the user and the target and the target is within range of the rangefinder. Lastly, if the user is interested in calculating the approximate elevation of the target, it's a simple matter of multiplying the slope distance by the sine of the slope angle (*Slope distance* × sin *(Slope angle)*) to derive the vertical distance between the user and the target feature. This distance can be added or subtracted (depends on whether the target is located above or below the user) to the elevation calculated by the GPS receiver at the user's position.

It is imperative to understand that augmented coordinates include the GPS position error plus errors related to the limitations of the other mapping tools and any human errors in measurement. Repeating measurements from the same user location can significantly reduce human errors and calculating augmented coordinates for the same target from different user locations can increase the accuracy of the derived location for the target. While both practical and simple, researchers should exercise great care when making these types of measurements and calculations and when training others to do so. Always take into account the sources of error and their potential impact on the results.

The previous examples illustrate how considerations of feature accessibility, feature shape, and data capture contingency plans translate into protocols in the field. Ideally, the GPS receiver should be located in direct contact or directly above the target feature when calculating and recording coordinates. When this is not possible, GPS coordinate measurements made at locations adjacent to the target feature and augmented GPS coordinates are acceptable provided they are calculated and uti-

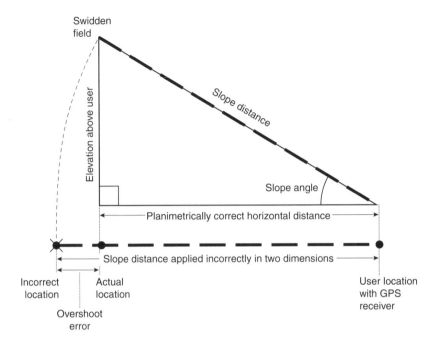

Figure 8.2. Slope distance vs. planimetric distance in augmented coordinate derivations

lized very carefully. Just as points provide the basic building block for higher dimension spatial objects, the various protocols used to locate point features are the basis for the data capture protocols used with higher dimension features.

Line features
Line features are target features that are represented in the spatial database as two or more coordinates with each pair of coordinates connected by a line to form linear spatial objects that have one dimension, length (Figure 8.1a). Examples include roads, paths, streams, rivers, canals, power lines, property lines, and transects. Line features are mapped using GPS receivers and applying one of three spatial data capture protocols to the real line feature, and then creating a line feature in the GIS by connecting each pair of successive coordinates with a line.

For line features that are accessible to the GPS user, two protocols exist. The first protocol involves recording coordinates at key points along the length of the line feature. If the line feature happens to be straight, the GPS user can stand at one end of the line feature, record a coordinate pair, move to the other end of the line feature and record a second coordinate pair. For line features that are not straight, it is necessary for the GPS user to record coordinates at key positions along the length of the line. A key position might include a curve, or a left or right turn along the length of a road or a bend in a river. To obtain greater detail in mapping a curve in a road or a very sinuous portion of a river, a greater, denser number of coordinates should be recorded along those sections of the line feature. In some applications, the key points along the length of a line feature may correspond to a change in some variable. For example, a researcher may establish a series of straight-line transects radiating out from a village center for a distance of 2 km. Each transect is traversed from beginning to end and coordinates are recorded at points along the transect where land ownership changes hands and land cover changes from one type to another. This method is the only option when using recreational receivers that only collect point locations. However, it is a desirable option for users of mapping grade receivers when the line feature is either very simple (e.g. straight) or very long, as this method requires less memory usage than the next protocol.

A second protocol for mapping line features using GPS receivers involves recording coordinates while traversing (e.g. walking, driving, biking) the length of a line feature. This is an option with those grade receivers that can collect rover files. The GPS receiver is configured to calculate and record coordinates at user-defined time or distance intervals. The GPS user should position himself/herself at one end of the line feature, turn the GPS receiver on, establish contact with the GPS satellites, open a rover file and record the rover file name, and traverse the length of the line feature as the receiver periodically calculates and records coordinates. At the end of the line feature, the rover file is closed and saved to the GPS receiver's internal memory. Consider this method carefully, as the frequency with which coordinates are recorded in a rover file has implications for how quickly the GPS receiver memory is filled, how frequently the GPS receiver will need to be downloaded, the length of time for downloading data, and the accuracy of representation of the linear feature.

A third protocol for mapping line features involves those line features that are not physically accessible to the GPS user. Provided that the relevant portion of the line feature is in view of the GPS user, augmented coordinates, as described previously, can be calculated for key points along the length of the line feature. The GPS user calculates coordinates for his/her position and then determines a look direction, look angle, and look angle distance to each key point along the length of the line feature. These inputs are used in a series of calculations to derive augmented coordinates for each key point. If more detail is desired in portions of the line feature that are more complex and sinuous, a greater number of key points will need to be identified along those portions of the line feature and augmented coordinates will need to be calculated for each of the key points. In all three protocols mentioned above, the resulting points in the GIS are connected with lines to form one-dimensional line objects.

Polygon features

Polygon features are target features that are represented in the spatial database by at least three coordinates connected by lines to form a two-dimensional spatial object that has area (Figure 8.1b). Examples of polygon features include agricultural fields, water bodies such as ponds, lakes, and reservoirs, community and village boundaries, forest boundaries, park boundaries, and political boundaries. Spatial data capture protocols for polygon features include different techniques for delineating the perimeters of polygon features. Protocols involve mapping key points in the geometry of the polygon feature and then connecting the points in the GIS software with lines in order to form the perimeter of the polygon.

If accessibility to a polygon feature is not an issue, two data capture protocols are available to the GPS user. The first option involves recording coordinates at the corners of the polygon feature. As with the similar method described for line features, this protocol is the only option for recreational receiver users, and is a desirable option for mapping grade receiver users when the perimeter of the area is simple or very long. A large rectangular agricultural field is a simple example. The GPS user needs only to position himself/herself at each corner of the agricultural field and record coordinates for each corner. Alternatively, the GPS user may wish to position himself/herself at some point on the edge of the agricultural field, turn the GPS receiver on, establish contact with the GPS satellites, open a rover file, and traverse the perimeter of the agricultural field as the GPS receiver periodically records coordinates in the file. Once again, this option is only available to users with receivers that collect rover files. After the GPS user has traversed the entire perimeter, the rover file should be closed and saved in the GPS internal memory. As with line features, this protocol also has implications for GPS receiver memory, data downloading frequency, time consumption, and accuracy of polygon representation.

A third protocol allows the GPS user to map the corners of a polygon feature remotely. This protocol is especially useful in areas of steep terrain and where accessibility to the polygon feature is limited. We revisit the swidden field example to help illustrate. Instead of capturing the approximate center point of the remote field, the researcher is interested in obtaining the dimensions of the field. The perimeter, but more importantly, the corners of the field, must be visible to the GPS user. The GPS user calculates coordinates for his/her position using the GPS receiver. Using a compass, clinometer, and rangefinder, the GPS user then determines the azimuth (look direction), look angle (slope), and distance (must be within the distance measuring capability of the rangefinder given the reflective properties of the target) along the look angle to each corner of the field. Inputting these measurements into the set of equations provided in Table 8.1 yields augmented coordinates for each corner of the field that are located at true horizontal (planimetric) distances from the GPS user's position. Thus, a planimetrically correct, two-dimensional polygon can be created in a GIS by connecting the corner points of the polygon using lines.

A fourth protocol utilizes the GPS receiver to record coordinates for a point location within a polygon feature of interest, such as a location within a forest stand, other vegetation patch or land use / land cover type. Information about the types of vegetation present or some other attributes about the site might also be recorded depending on the nature of the research. This combination of GPS data and the description of land use / land cover surrounding the GPS location is called ground truth data. This procedure is typically repeated for a number of distinct areas in a study region in order to aid in classifying a satellite image of the study region into land use / land cover types. The coordinates are used in conjunction with georeferenced, digital aerial photographs and/or georeferenced satellite imagery in order to identify the same polygon areas (e.g. forest stands, vegetation patches) in the photos or imagery that are coincident with the GPS point locations. Ideally, the aerial photographs and satellite imagery should be acquired around the same time that the GPS coordinate data were recorded in the field. With the GPS point locations overlaid on the aerial photography and/or satellite imagery, it is possible to delineate the boundaries of the coincident polygon features using manual digitizing. The collected attributes can then be attached to the polygon objects in the GIS.

Three-dimensional features
Three-dimensional (3-D) target features are represented in the spatial database as horizontal coordinates (x,y locations) with one or more z-values

associated with each coordinate pair. The z-value typically represents some value, measurement, or observation at each location. *Surfaces* (Figure 8.1c) usually have one z-value, such as elevation, associated with each of at least three locations. On the other hand, three-dimensional *volume objects* typically have more than one z-value attached to each of at least three locations. The result is a spatial object that has volume (Figure 8.1d) with dimensions of length, breadth, and depth that is also geographically referenced. Collecting 3-D feature data using GPS receivers is less common in social science research compared to the point, line, and polygon feature types described above, but it does warrant attention as it is a viable means of collecting spatial data for more sophisticated mapping and modeling purposes and because 3-D modeling capabilities and functionality are becoming more common and user friendly in the GIS environment.

Most GPS receivers use distances to at least four satellites in order to calculate a three-dimensional position, which includes horizontal coordinates and a vertical position, or elevation (altitude), equivalent to the height above the WGS84 ellipsoid (commonly abbreviated as HAE). GPS users should note that HAE is different than the height above mean sea level (MSL; see Chapter 4) and should consult a professional surveyor or other reliable source for determining the difference between the two measures for a particular area. The accuracy of the HAE varies due to the same error sources that affect horizontal accuracy (see Chapter 5). In general, one can expect the accuracy of raw HAE measurements to be off anywhere from a few meters to tens of meters unless high accuracy error correction methods are utilized in the work. If the receiver has the capability, one way to improve the accuracy of raw HAE measurements is to calibrate it by entering the known HAE (measured or obtained by other means) for a particular location in the vicinity of the study area. GPS error correction techniques, as discussed in Chapter 5, can also improve the accuracy of the HAE.

Typical GPS data capture protocols for creating an elevation surface for an area might include the calibration just mentioned and then occupying the desired locations within the area and recording GPS data in 3-D mode to ensure that a latitude, longitude, and altitude are recorded at each location. Users are strongly encouraged to consult the GPS receiver documentation and with GPS vendors and survey professionals if interested in creating a detailed elevation surface using GPS technology. If the GPS user is interested in creating a surface using some other attribute besides elevation, the protocol is the same, except the alternative attributes will replace the altitude value in the spatial database. For example, the GPS receiver might be used to record the geographic coordinates for soil sample locations. One or more soil attributes, obtained from analyzing the soil samples, can be attached to the sample locations and used to create a surface based on the values for each soil property at each location.

Three-dimensional volume objects differ from surfaces in that for each horizontal coordinate pair there are typically two elevation values. One elevation value is found at the bottom of the volume object and one elevation value defines the top of the volume object above the other elevation value. At least three horizontal coordinates are needed in order to create a column of space. For example, public health officials announce that several species of birds have been shown to be particularly sensitive to the West Nile Virus with most of these species dying within one to two weeks of exposure to the disease. Knowing birds have always been important sentinels for the general health of the environment, an epidemiologist and biologist coordinate an effort to monitor and evaluate several of these bird species in their forest habitats located near three large suburban areas. Several intensive monitoring sites are established and the researchers wish to map the 3-D extent of the forest habitat at these sites as well as the range for these species of bird in relation to adjacent human habitat and any potential mosquito breeding sites. By doing so the researchers hope to identify the presence of the disease early on and take measures to prevent the spread of the disease to the nearby human population.

Given that the tree canopy cover is quite thick in the forests, the researchers might use an external antenna that is capable of extending above the tree canopy. Alternatively, the researchers might be able to position themselves and the GPS receiver with antenna above the tree canopy. The GPS receiver is then used to calculate a geographic position that includes an elevation for the position of the antenna above the canopy. By measuring the height of the antenna above the ground (using a rangefinder or meter tape), a second geographic position with elevation can be derived for a location on the ground directly below the measurement made above the canopy. Following this same type of protocol for a number of locations around the edge of the birds' forest habitats, the data can be used to model three-dimensional, volumetric habitat spaces within a GIS. If canopy cover is not an issue, then the GPS receiver can be used to determine a geographic position with elevation at ground level. Calculating the above ground elevation at each position can be done using an extendable external antenna, positioning the GPS receiver with antenna at the above ground position or measuring the distance to the above ground position from the ground level position. Just as with surfaces, the GPS user should take care to understand the accuracy and limitations of the specific GPS receiver in measuring elevation and any calibrations to the known altitude or antenna height that are necessary.

Feature attribute data

Attribute data that are spatially linked to geographic features facilitate powerful and insightful mapping and spatial analysis. As such, there are some basic considerations regarding their linkage and content that must be made during the planning stage. One of the benefits of using GPS receivers to collect spatial data for GIS work is that most GPS receivers automatically assign unique identifiers to each coordinate pair or spatial data file collected, usually in the form of waypoint names or rover file names. These names become a basic, unique attribute tied to a specific coordinate pair or data file. The uniqueness of these names is important, as the names are typically used as key identifiers for each of the coordinates or data files and their other associated attributes throughout the remainder of the research.

GPS rover file names/waypoint names and corresponding coordinates should always be recorded as initial feature attributes on a hardcopy data form, or some other hardcopy point log such as a field notebook, or directly on a survey questionnaire being completed for a target feature that is being located. This provides a hardcopy backup of the GPS coordinate data saved in the internal memory of the GPS receiver. In the event that the GPS receiver malfunctions and data are lost, or if data are inadvertently deleted from the GPS receiver memory before downloading, the hardcopy log can be referenced to obtain the missing spatial attributes.

Additional attributes should be collected for each target feature based on the information that is needed about that target feature in order to answer, or attempt to answer, the research questions. A list of these attributes and the possible values that might be encountered in the field for each attribute should be created and used to develop the project's GIS spatial database and any data collection forms (e.g. hardcopy forms, electronic forms, digital data dictionaries, field notebooks) used in the fieldwork. Feature attribute data may be gathered at the same time as the spatial data or in a separate effort. For example, survey questionnaires developed and/or administered independently of spatial data collection can be a source of feature attribute data provided there is a link between the spatial and survey questionnaire data, via a unique, key identifier for each target feature (e.g. household, village, facility, etc.) that is common to both data sources.

Summary

Decisions regarding the selection, description, and graphic symbolization of features and phenomena should be made after careful consideration of the research questions, the nature and scales of analyses, and the available resources for carrying out fieldwork. In terms of data, the researcher needs to ask what will be located, how will it be described, and how will it be symbolized. By answering these questions, the characteristics of the spatial data and a spatial database schema are realized. But how do these answers translate into a data collection methodology? The second part of this chapter addressed this question by outlining the development of methods to collect the spatial data necessary to transform the real features and phenomena of interest into spatial objects with attributes in the GIS.

Knowing what data to collect and how to collect it is only the first step towards successful data collection. As with any work environment, knowledge *and* resources are needed in order to produce results. The following chapter discusses field resources, specifically, equipment, personnel and field instruments, which comprise the second component of fieldwork planning and preparations.

9
Fieldwork Planning and Preparations: Field Resources

Introduction

As the second component of fieldwork planning and preparations, this chapter consists of a broad, annotated listing of key field resources applicable to a variety of projects and field situations. Equipment, personnel, and field instruments represent significant monetary investments in the data collection effort, thus it is essential to consider the "what" and "whom" needed in order to complete the research. The chapter is organized into three corresponding sections: common field equipment, personnel, and field instruments. The lists presented below are by no means exhaustive, and given the variation of specific project needs, some or many aspects may not pertain.

Common Field Equipment

This section includes brief descriptions and general uses for common field equipment. It is very important to understand the capabilities and limitations of all field equipment before purchasing and training others in how to use the equipment. Consult with vendors, knowledgeable users, and refer to the specific product user manuals for specifications, capabilities, limitations, and detailed instructions for use.

Equipment prices vary by vendor and may be drastically different by the time of the printing of this book. Therefore, we have not included specific product prices here. Instead, we suggest collecting bids from several vendors for big-ticket items, such as the GPS receivers, hardware, software, and other electronics. Many vendors offer educational, government, volume, or other special discounts. For smaller items, shopping around locally or on-line will suffice. Once in the field, certain items, such as batteries and film, may need frequent replacement. Shopping around within the study area, or in nearby towns and cities, is suggested.

GPS receivers

When selecting a GPS receiver for fieldwork, it is important to first have a clear understanding of the application of the GPS technology, which leads to the selection of an appropriate error correction method that will provide the level of accuracy required. For most social science applications, good quality recreational receivers can provide adequate accuracy. Most are relatively inexpensive, easy to use and can be found in many large department stores and outdoor recreation stores. Mapping grade GPS receivers are recommended for projects requiring high accuracy data. Most importantly, one should acquire GPS receivers capable of the error correction method and accuracy desired. Describing the proposed applications to a GPS vendor or an experienced user will help determine the type and number of GPS receivers required. For a more thorough discussion of the types of GPS receivers and the important features and capabilities to look for in a receiver, please refer to Chapter 5.

GPS receiver accessories

Apart from the features and capabilities of the receiver, it is important to understand what is included with the receiver. Does the purchase price include the receiver alone or an entire GPS field package? Some vendors offer packages that include two receivers. GPS operation manuals should come standard with the receiver. In addition to manuals, the user should inquire about other items such as a carrying case, external antenna and antenna pole, power supplies (e.g. AA alkaline and rechargeable batteries, 12-volt camcorder rechargeable battery, vehicle power adapter), data transfer cables, memory cards, and spare parts. Most mapping grade receivers and some recreational receivers include a port for connecting a compatible external antenna. A telescoping paint pole is an inexpensive alternative to the frequently overpriced antenna range poles available from GPS vendors. Other antenna mounting options include the top of a vehicle, motorcycle and bicycle mounts, tripods and backpacks, or even tree limbs and bamboo.

GPS software

All mapping grade and some recreational grade GPS receivers have a connection port that permits the transfer of data from the receiver's internal memory directly to a PC or data logger. In addition to requiring an interface cable and a computer (desktop, laptop, hand-held computer, palm-top computer), it is necessary to have application software that can interpret the data stream from the GPS receiver and output the stored coordinates and any attribute data to a computer file. Some receivers, usually the higher-end mapping grade receivers, may ship with their own software. Other GPS receivers require users to purchase the software separately. Fortunately, many independent GPS users have developed software packages, extensions, and add-on modules that are specifically designed for downloading GPS data, are available for little or no cost over the internet, and work in conjunction with many existing GIS software packages.

As with the GPS receivers, different software packages provide a variety of different features. At the high end of the spectrum, the software can (1) aid in fieldwork planning by determining the optimum time to be in the field based on satellite orbits and geometry, (2) perform post-processing differential corrections, and (3) output the results to a variety of file formats readable by GIS software. The more basic software packages may only take the data stored in the receiver and save it to a text file, which will require additional processing before it is ready for use in a GIS. Many GIS/GPS extensions and add-on modules claim to transfer raw and real-time differential data from a receiver directly to the GIS or image-processing environment. Post-processing differential correction usually requires a standalone GPS software package. Users should first determine if GPS software is needed for their particular GPS application. If software is needed, the user should compare the features, functionality and receiver compatibility of various software packages and actually test the software with the GPS receiver prior to making a purchase or gathering real project data.

Computer and accessories

A computer and proper software are needed in order to download and process the data recorded by most GPS receivers. Laptops, tablet PCs, and some hand-held computers can be used to download data in the field. These same computers can be used for storing and processing the data as well. Desktop PCs may be used when downloading and processing of data can wait until the return to the office. It is, of course, important to acquire a computer that meets the minimum system requirements suggested by the GPS software. Some GPS receivers are essentially hand-held computers themselves, with the capability to run limited GIS and mapping software, perform file management, as well as perform the full suite of GPS receiver functions.

PC interface cables and/or memory cards can be used to transfer data from a receiver to a computer. Power supplies should include the computer's rechargeable battery and a reliable AC energy source for recharging the computer battery. Most laptops come with an AC power adapter with voltage input ranging from 100 to 240 volts. For data backup purposes, a good supply of floppy disks, writeable or rewriteable CDs or DVDs, or other storage media should be acquired. A durable computer carrying case is also recommended, especially if the hardware will be taken into the field.

Other mapping tools

Some social science projects require not only the coordinates of target features, using a GPS receiver, but also angle and distance measurements associated with the target feature and other measurements and observations that require more specialized equipment, sensors, gauges, etc. Measuring angles and distances for navigation and mapping purposes can be done using several relatively inexpensive analog and digital tools, including a map compass, clinometer, rangefinder, and meter tape. Prices for these items vary by brand and quality, but most can be purchased for under US$300.00 each. Most, if not all, can be researched and purchased at survey supply shops and outdoor recreation stores and their websites. See Boxes 8.1–8.4 for a description of commonly used mapping tools.

Reference maps

Although a GPS receiver can determine a location on the Earth's surface fairly accurately, it is important to have the geographic context within which to identify GPS derived coordinates and the GPS user's position relative to the surrounding environment. Reference maps provide this context and should be acquired and/or produced for the study region as part of the field-work preparations. They can be used as a planning tool, and data collection teams can use the maps to familiarize themselves with the study area and identify and mark potential areas or sites to visit. Once in the field, reference maps become navigation tools, which can be used in conjunction with the GPS receiver and map compass to find the best route to the next site or to some other location. Maps can also serve as data recording devices; locations that have been visited and/or surveyed can be marked or delineated on reference maps and information, such as site ID and other attributes, can be written directly on the maps.

Field preparations should include obtaining map coverage of the study area, if possible. Maps of various types and scales may be used in order to achieve full coverage, however consistency and sufficient detail among them is encouraged. For some projects, the study area may be so small, as in a household survey of a small village or housing development, and/or so familiar to the data collectors, that detailed maps are unnecessary. At a minimum in this case, a house roster should be created and used to keep track of data collection progress and to ensure that no households are overlooked. For larger, unfamiliar and/or remote areas, reference maps become an invaluable necessity. Lastly, maps should be protected from the elements as much as possible. Protective folders, plastic cover sheets, waterproof map tubes, and map lamination are some available options.

Cameras and accessories

Some projects may require data collection and documentation in the form of photographs and/or video footage. For example, using GPS receivers to ground truth satellite imagery typically involves taking one or more photographs and/or recording video footage at each GPS location in order to document the ground cover and land use. A large variety of good quality cameras are available on the market for under US$500.00.

When using traditional cameras, always have a sufficient supply of film on hand and consider the additional costs of developing and scanning the photographs. For digital cameras, additional memory can be purchased in the form of various types of memory cards to expand their standard capacity. The number of pictures that can be stored depends on the total amount of memory and the image quality setting on the camera. When a memory card is full, it can be switched out with an empty card, or it can be erased after its contents have be downloaded. This is achieved using an interface cable that connects the camera data port to the computer. More expensive digital cameras can also record motion pictures, but this usually requires a significant amount of data storage space. An additional benefit of digital cameras is that some GPS manufacturers now have hardware and/or software that can link GPS data to a digital photo. These applications range from linking the photo file and the coordinates in a software package, to physically connecting the GPS receiver to the camera and having the coordinates imprinted on the photo itself.

Using camcorders for recording video requires a good supply of tapes of the relevant format. In addition to the standard analog formats (VHS, 8mm, etc.), new digital camcorders are available which utilize digital tape formats (Hi8, MiniDV, etc.) to provide better video quality. Sufficient power sources, such as AA alkaline or rechargeable batteries, and/or camcorder batteries should be acquired in addition to a good quality carrying case. Special accessory needs, such as zoom and macro lenses, tripods, etc. should also be considered during fieldwork planning.

Communication devices

For reasons of coordination, safety, and security, reliable communication devices are highly recommended, especially for projects involving multiple data collection teams, limited transportation options and/or data collection in remote areas. Two-way radios are suggested for remote areas where cellular mobile phone service is unavailable. Relatively inexpensive two-way radios with approximately a 2 mi range, depending on terrain and other local conditions, and AA battery power source can be purchased at many department stores and outdoor recreation stores. Some manufacturers make a combination GPS receiver and two-way radio. More powerful two-way radios can be used, but they usually require special FCC licensing permits and associated fees when purchased in the United States. Similar systems exist in other countries. Digital satellite phones and services can be costly, but communication range is virtually unlimited worldwide. For study areas located within and in the vicinity of most urban areas and major traffic corridors, the benefits of cellular mobile phone service can be enjoyed, whereas less developed countries may have problematic, spotty service, or none at all. Also, be mindful of phone battery consumption and recharging times when planning.

Miscellaneous items

Field notebooks, binders, paper, pens, pencils, and markers should be in good supply in order to take notes, record data, sketch maps at sites, and label bags and equipment. Some notebooks and pens allow data collection in the rain. These should be considered for wet climates. Plastic kitchen storage bags are recommended for storing smaller items and keeping equipment dry. A set of tools and duct tape, electrical tape, and glue are recommended for fieldwork in remote areas. These items are used for making basic repairs in the field, such as wrapping a cut in a cable with electrical tape. It is also a good idea when working in remote areas to have extra data download cables and other replacement parts on hand. While it is impossible to provide a generic comprehensive list of these items, anticipating potential problems by having spare parts available can help prevent lost time.

Personal equipment

Each member of a data collection team requires a certain number of personal items and supplies. The duration and location of the fieldwork should be considered when creating a personal equipment checklist. Comfortable clothing that is appropriate for the range of temperature and weather associated with the study area should be packed in addition to appropriate shoes for the field situation. Backpacks or some other comfortable carrying case should be used for transporting supplies. Medicines (disease preventative, such as antimalarial, and personal medications), insect repellent, sunscreen, sunglasses, hats, and personal emergency first aid kits should be packed as needed. For outdoor overnight stays, flashlights, tents, and sleeping bags are recommended.

Personnel

Apart from principal researchers, several key field personnel are recommended for participation in the data collection effort. As with equipment, the type and number of personnel required will vary from project to project based on research objectives, accuracy needs, and the size and intensity of the undertaking. Key field personnel should be familiarized with a project's goals and objectives and understand how the data collection effort fits into the overall project. Projects with a local or regional study area may only require personnel recruited from or near the researcher's home institution. National and international study sites located at great distances from the home institution will benefit greatly by employing personnel from on- or near-site collaborating institutions, particularly in areas where safe navigation, language barriers, and/or political and cultural sensitivities and infrastructures are concerns.

   Key personnel roles are described in the following paragraphs. Descriptions include suggestions for basic knowledge, preferred qualities, and responsibilities. Personnel salaries and hourly rates are not discussed as they are subjectively determined in many cases, may fluctuate with the economic conditions in the study region, and are subject to varying project budgets. It is important to note that a single person may be able to function in multiple roles. Moreover, multiple persons may be required to fill certain positions. Again, it depends on the scope and nature of the project and the data collection effort.

Project manager

This person should be familiar with all aspects of the project, including goals and objectives, budget, personnel, time lines, data collection and integration methods, and survey instruments (spatial and nonspatial). The project manager works closely with the principal researchers, the collaborating institution's investigators and officials, the spatial information specialist, and the data collection team leaders. While a working knowledge of GPS is not necessary, a basic knowledge of the GPS system and the reasoning behind the spatial data collection and its integration with other types of spatial and aspatial data is recommended. Obviously, organizational and interpersonal skills are a must for the project manager.

Spatial information specialist

The spatial information specialist should possess an advanced knowledge of GPS, GIS and spatial data collection, management, integration, and analyses. He/she works closely with the project manager, field technicians, data collection team, and local area experts. He/she is responsible for developing spatial data collection protocols and field instruments; training the field team in GPS use/maintenance, spatial data collection protocols, and proper use of the field instruments; ensuring that functional GPS receivers and other field equipment, including reference maps, are available and maintained; ensuring that spatial data collection protocols are understood and applied consistently throughout the fieldwork; and seeing that GPS data and documentation are accurately maintained and returned to the home institution. The spatial information specialist should take special care in training one or more field technicians, establishing criteria for selecting knowledgeable local area informants/experts, and conducting any necessary training of the local area experts.

Field technician

The field technician should possess an intermediate to advanced knowledge of the GPS system and a working knowledge of all field equipment, software, and computer-related data management tools. The field technician typically becomes the spatial information expert in the absence of the spatial information specialist, and should be trained to troubleshoot equipment problems encountered in the field and be able to train others in GPS use and equipment troubleshooting. This person is directly responsible for periodically downloading and archiving GPS data and maintaining accurate data documentation. The field technicians and the home institution's spatial information specialist should work closely together and communicate often throughout the data collection process.

Data collection team

One or more data collection teams consisting of one or more persons each may be used depending on the number of target features to locate, the size and complexity of the study area, and the length and intensity of the fieldwork. Members of the data collection team should possess a basic knowledge of GPS and a working knowledge of the equipment and spatial data collection protocols. They will be responsible for the data collection forms, maps, and field equipment that are assigned to them, and they might be asked to carry money for hiring local informants as needed during the data collection phase. They should be able to effectively manage the data they collect, including checking and correcting their data, and indicating where and why corrections were made when the data is returned to a central location, such as project headquarters. Each data collection team should have a team leader appointed who can organize and manage the team, get feedback from the team regarding working conditions and protocol suggestions, serve as the team spokesperson, ensure that data collection protocols and checks are being performed properly by the team, and provide moral support to the team.

Data collection team members may be recruited from collaborating institutions, the home institution, and/or other sources the researcher deems appropriate. Depending on the setting, having local team members and researchers may lessen unease among the local population and facilitate cooperation with the project. Therefore, if possible and where applicable, recruiting and training a data collection team that originates from, or in the vicinity of, the study area is encouraged. Doing so can greatly benefit a project by having team members who can speak the local dialects, are familiar with local customs and the local cultural, political, socioeconomic and environmental conditions, and who are more likely to earn the trust of local respondents, allaying any fears that respondents may have and allowing them to respond more freely. Obtaining assistance from the local communities with the data collection is mutually beneficial to the research project and the communities. A research project can empower local communities by imparting knowledge and skills and providing opportunities that would otherwise not exist.

Local area expert

The local area expert should be familiar with many of the geographic, political, social, cultural, economic, and historical aspects of the study area. Ideally, the local area expert should originate from the study area, speak the local language / dialects, and interact well with local and regional inhabitants and officials. In many ways, the local area expert serves as a project diplomat, facilitator, and guide, who can explain a project's goals, objectives, and data collection protocols to the local and regional population; assist in the data collection and in obtaining any sensitive information, documentation, and permits; safely lead researchers throughout the study area, providing their own unique insights and narratives about the study area; and perform other tasks as needed. Local area experts might also be able to make other local inhabitants feel more comfortable with the team and the project. Lastly, without the knowledge and aid of local area experts, vital information about the study area or study sites could be overlooked and lost.

Field Instruments

This section outlines the various field instruments that should be considered and designed for use in fieldwork. Unlike field equipment, field instruments in this context are not devices one takes into the field to make measurements or take readings. Rather, they are the collection of documents used by the data collection team to query, derive, record, and organize information and they facilitate the data collection process. In addition to listing and briefly describing some of the documents that may be used, some sample data collection forms are provided. As projects vary, so do the field instruments that may be required. The type of research and specific project needs dictate the contents of field instruments, however, they can be classified into three general types, including data collection forms, reference maps and graphics, and reference lists.

Data collection forms

Data collection forms are the field instruments used to record information about a particular target feature. They include hardcopy and electronic data forms specifically designed for spatial and attribute data collection; non-spatial survey questionnaires with an area on the questionnaire for entering spatial information; digital data dictionaries loaded on a GPS receiver, hand-held PC, or some other computing device; and field notebooks. When designing forms that will include GPS data collection or when recording information regarding GPS data collection in a field notebook, there are a number of common fields to include on the form or record in the notebook that, if completed properly during fieldwork, result in a full record of the GPS data collection process and document some of the fundamental attributes about the target feature being located. Figure 9.1 specifies these suggested fields for better GPS data collection.

The *Site Name* or *Site ID* can be a basic, unique attribute describing each of the field sites. Also included on the form is the *Date* of the data collection and the GPS receiver *Serial Number*. The serial number is useful when using more than one GPS receiver during the fieldwork. A *Site Comments* field can be used to record any anomalies, variability, or other interesting observations about the site. The *Team Members* fields are used to record the names of the data collectors using the GPS receiver. The *GPS Waypoint / Rover File Name* field is used to record the name of the waypoint or rover file assigned to the coordinates or geographic data file saved to the internal memory of the GPS receiver. The waypoint or rover file name can be used as the key identifier that links the spatial object created from the coordinates and/or GPS data file with the other attribute information contained in the data collection form. These names should follow predefined file naming conventions established by the research team such that a unique file name is created for each waypoint and rover file collected. Depending on the nature of the research, *Elevation*, or height above the ellipsoid, may be included on the data collection form. *Waypoint X* and *Waypoint Y* fields are used to record coordinates for the site. These coordinates serve as a hardcopy backup for the waypoint or rover file

Site Name: ....   Date: ....

Site ID: ....   GPS serial #: ....

Comments on Site: ....

Team Members: 1) ....   2) ....

GPS Waypoint or Rover File Name: ....   Elevation (if applicable): ....

Waypoint X: ....   Waypoint Y: ....

Comments on GPS Receiver and/or Positioning Method: ....

....

Sketch Map (if necessary):

Additional attribute data fields and/or preexisting data forms inserted here:

Figure 9.1. Suggested fields to use when designing forms incorporating GPS data collection

recorded in the GPS receiver memory. It's important to note that a rover file collected with a mapping grade receiver does not inherently have one coordinate pair assigned to it. Rather, it is a collection of numerous individual coordinates. Therefore, it is recommended that a point-averaged waypoint be collected in addition to any rover files, and recorded on a field instrument, to serve as a backup. In lieu of having two separate receivers, this can be accomplished by capturing both files sequentially with the same receiver. A field for comments on GPS receiver and error correction method is used to describe any special circumstances under which the waypoint/rover file was collected or to describe any changes to the GPS configuration that were necessary or to record similar notes. If deemed necessary, an area should be designated on the data collection form for drawing a sketch map of the target feature, its surroundings, and the locations of GPS measurements of the target feature. See Appendix B for an example of a generic, multipurpose GPS field instrument.

Additional attribute fields can be added to these common GPS-related fields in order to customize the data collection form to suit the needs of a project and the target features under study. If a project's data collection form already exists, the researcher can simply add the GPS-related fields into the existing form. See Figure 9.2 for an example of a data collection form for gathering GPS ground control point data, using geodetic coordinates at road intersections in order to orthorectify a series of aerial photographs. The basic GPS-related attribute fields are built into the form, including an area for sketching maps of the road intersections in order to identify exactly where a GPS measurement was taken at each intersection. Additional attribute fields inquire about the photographs being rectified, the number of roads converging at the intersection, and the road surface types found. The form in Figure 9.2 merges GPS data collection with another type of spatial data, aerial photographs.

GPS-related attribute fields can also be incorporated with data collection that was previously non-spatial in nature. Figure 9.3 illustrates the first page of a sample data collection form for recording locations and selected attributes of health facilities organized by city/town and county within a study region. The basic GPS-related fields are included in the form as well as city, town, and county name and ID and a series of fields related to facility characteristics. The GPS-related fields could be stripped out of this survey form and a valid data collection form for gathering both non-spatial and relative location characteristics (city/town/county/street address) of health facilities would remain. This form could have been fit with GPS-related fields after it had been previously used without GPS in mind, or the form may have been designed from the start with GPS data collection in mind. Either way, adding a GPS component to new or existing data collection forms is easy to do, and adding the spatial component can prove invaluable.

Figure 9.3 shows an example of a simple data dictionary. The Facility

Geodetic Control GPS Data Collection Form

Team members: ....   Date: ....
....   GPS serial #: ....

Site ID	GPS information	Photo ID(s)	Photo date(s)	Photo quadrant	Feature description	Drawing / sketch
	File name: .... Antenna ht.: .... Latitude: .... Longitude: .... Elevation: .... Method: .... Comments:				.... 3-way .... 4-way .... dirt .... concrete .... gravel .... asphalt .... other	
	File name: .... Antenna ht.: .... Latitude: .... Longitude: .... Elevation: .... Method: .... Comments:				.... 3-way .... 4-way .... dirt .... concrete .... gravel .... asphalt .... other	
	File name: .... Antenna ht.: .... Latitude: .... Longitude: .... Elevation: .... Method: .... Comments:				.... 3-way .... 4-way .... dirt .... concrete .... gravel .... asphalt .... other	
	File name: .... Antenna ht.: .... Latitude: .... Longitude: .... Elevation: .... Method: .... Comments:				.... 3-way .... 4-way .... dirt .... concrete .... gravel .... asphalt .... other	
	File name: .... Antenna ht.: .... Latitude: .... Longitude: .... Elevation: .... Method: .... Comments:				.... 3-way .... 4-way .... dirt .... concrete .... gravel .... asphalt .... other	

Figure 9.2. Example of a ground control data collection form for a series of aerial photographs

Codes section of the form indicates the attributes associated with each facility location. For some of the attributes, the possible values that the user can assign the attribute are listed. Limiting responses to a predefined set of values ensures some level of objectivity in the data collection, but it can also be problematic if a target feature cannot be properly described by the predefined set of values. If this occurs, the available information should be recorded and a note should be made as to why the target feature fell outside

Health Facility Data Collection Form

Date: .... Start Time: .... End Time: ....

City/Town Name: .... City/Town ID: ....

County Name: .... County ID: ....

Team Members 1) .... 2) ....

GPS serial Number: .... Comments: ....

Facility Codes:

TF (Type of facility): 1. Private Hospital 2. Public Hospital 3. Outpatient 4. HMO 5. Other
LC (Level of care): 1. Primary 2. Secondary 3. Emergency 4. Other
ICU (Intensive care unit): 1 = Yes; 2 = No   #MD = Number of medical doctors
PH (Pharmacy): 1= Yes; 2 = No   #RN = Number of registered nurses
FPS (Family planning services): 1 = Yes; 2 = No   #BD = Number of beds

GPS file name	GPS display coordinates	Facility info	Facility respondent	T F	L C	L C U	P H	F P S	#MD	#RN	#BD
	Latitude	Name	Name								
	Longitude	Address	Position								
	Latitude	Name	Name								
	Longitude	Address	Position								
	Latitude	Name	Name								
	Longitude	Address	Position								
	Latitude	Name	Name								
	Longitude	Address	Position								

Figure 9.3. Example of a data collection form for recording health facility locations and selected attributes in cities and towns within a study area

the range of predefined values. The range of values or the attribute itself may need to be reevaluated.

Data dictionaries (discussed in Chapter 5 and later in Chapter 10) can also be created for direct download to and use in some GPS receivers, usually mapping grade GPS receivers or better. To the extent possible, data dictionaries and other types of hardcopy GPS data collection forms used to collect spatial and attribute data about target features should be a reflection of the GIS spatial database structure. Customized digital data dictionaries are a prime example of how data collection forms can be created to match the structure of a GIS database. Table 9.1 shows an example of a community feature in a data dictionary. For each community target feature in column one, the geographic location is obtained, either as a waypoint or in a rover file, and saved as part of that community feature. In addition, there exists a number of feature attributes in column two of the data dictionary that must be assigned attribute values from those values available in column three. Some attributes require a numeric value, such as the total population and the number of households in the community. The remaining feature attributes are assigned values based on the available responses listed in the attribute value column. When all data are entered and saved, the community feature is closed and the GPS user may move on to another task or to the next community.

Table 9.1 Example of a simple data dictionary for describing selected attributes of rural communities in a study region

Feature	Feature attribute	Attribute value
Community – the feature that is being located	Total population	Numeric value
	Number of households	Numeric value
	Primary dialect – the dialect spoken most often by the people in the community being located	Dialect 1 Dialect 2 Dialect 3 Other (specify)
	Secondary dialect - the second most used dialect in the community being located	Dialect 1 Dialect 2 Dialect 3 Other (specify)
	Primary cash crop – the cash crop grown by most of the farmers in the community being located	Paddy rice (variety 1) Paddy rice (variety 2) Corn Other (specify)
	Plowing method – the plowing method used by most farmers in the community being located	Ox Mechanized Other (specify)

Field maps and other graphics

Field maps include general reference maps and maps specifically designed for recording or obtaining data. Both types of maps can be used for navigation and are most useful when a valid coordinate system and datum are used and a grid or set of tic marks in the proper map units is displayed on the map.

Field instrument maps are specifically designed for recording or obtaining data. They should contain the project's established coordinate system, datum and relevant map units. For recording purposes, maps can be drawn upon in order to identify or delineate the approximate locations and/or boundaries of certain features. Information about the features can be written on the map as well. This spatial and attribute information can be used as input into a GIS (with the understanding that the locations are approximate and contain error) or used in planning subsequent fieldwork, such as dividing up areas among data collection teams or identifying areas that data collection teams should avoid. Field instrument maps may have satellite imagery, land use or land cover information, aerial photography, elevation, or some combination of GIS thematic layers displayed on them. GPS data can be used to validate and/or correct the spatial and attribute information drawn or displayed on field instrument maps.

Field instrument maps may also be used for obtaining data. For example, a public health researcher may carry a detailed thematic map of soil arsenic levels for a county into the field in addition to a GPS receiver. Using locations obtained from the GPS receiver for particular sites of interest, the arsenic levels coinciding with the GPS locations may be read directly from the map in the field and immediately recorded in a data collection form in a record for each location. Thus, depending on the research project, field instrument maps can serve as data recording devices and/or data sources.

Other graphics might also need to be prepared for data collection teams and respondents to reference in order to record information about a target feature or to validate the presence of a certain phenomenon. For example, a project using GPS receivers to locate the occurrences of certain bird species might include pictures and descriptions used to identify those birds of interest. An epidemiologist interested in mapping the occurrences of certain insect-borne diseases might create graphics and descriptions of skin lesions and other symptoms associated with the diseases. A data collection team could then reference this information during village interviews, validating the presence of the skin lesions among villagers, and then mapping the village location or perimeter using a GPS receiver. A biogeographer may want to create graphic visualizations of percent cover classes to aid in describing the tree canopy cover or plant density in plant communities that are being mapped using GPS receivers. An anthropologist evaluating the spatial distribution of affluence across a large archeological site might carry photographs and detailed descriptions of various household items that are historically representative of economic status. The relevant artifacts can be identified using these reference materials and their locations can then be mapped using GPS receivers. Similarly, an archaeological team investigating agricultural land practices on a nineteenth-century plantation might benefit from pictures and descriptions of soil profiles and properties in mapping the locations of and recording attributes about soil pits dug within the study area. These are but a few examples of the types of reference graphics and supporting materials that may need to be carried by data collection teams in the field. These materials should also be used in training the data collection teams.

Reference lists

Reference lists refer to the various lists that research team members may need in order to keep track of equipment and maintenance; refer to in order to obtain, validate, and cross-check data; and those used to keep track of tasks and survey progress. An equipment checklist is useful for those who are responsible for managing a large amount of equipment and ensuring that the equipment is working properly through regular maintenance and a ready supply of replacement parts and exhaustible items (e.g. ink, paper, batteries, etc.), depending on the equipment.

Data-related lists might include codebooks or tables that the data collection team references in order to replace certain survey responses with a project-related string or numeric code. This may be done in order to transform responses into more easily analyzed forms or for data confidentiality reasons. Other lists may include acceptable values or ranges of values for certain survey responses. These types of lists are frequently used for data validation and cross-checking.

Process lists might include a list of tasks that a data collection team must perform upon reaching each target feature. The data collection team is instructed to check off each task as they complete it to ensure that each target feature is completely and consistently surveyed. These types of process lists may also be created for downloading and data backup activities. If the names of target features and their IDs are determined and assembled into a list prior to the data collection commencing, each data collection team should obtain a list of the target features they are responsible for locating and surveying. Examples include lists of households, villages, towns, and facilities. Teams can check off target features from the list as they complete the GPS and survey data collection for each feature, and these checklists can be used to keep track of the progress of the data collection teams.

Summary

Depending on the research project, the range of field resources outlined in this chapter may be overkill, grossly lacking, or somewhere in between. Researchers should determine where their project fits in relation to these lists and make adjustments as needed. After the necessary field equipment has been acquired, items such as the GPS receiver and antenna, computer, and other electronic devices should be assembled, configured, and tested prior to entering the field, following the instructions in the user manuals. These steps are also recommended just after the arrival of the equipment in the field. Specifics regarding equipment configuration and operation, particularly that of the GPS receiver and related hardware, are discussed in Chapter 11. Food and medical supplies are important to consider during fieldwork planning as well. They are addressed in the following chapters.

In terms of personnel, some projects may be of a size requiring only a single person to serve as the project manager, spatial information specialist, field technician, and data collector, whereas other projects may involve a team of 100 individuals. Regardless of the size of the team, it is important to assemble the human resources with the mixture of skills necessary to successfully carry out the data collection. For efforts involving GPS and spatial data collection, several specialized personnel roles are suggested. It is also a good idea to have backup personnel for some of these positions, if possible. Ultimately, personnel are only as good as their relevant experience, the other resources they have to work with, and the training that they receive.

Other resources include not only the field equipment discussed earlier, but also the field instruments, such as survey questionnaires, data collection forms/logs, and data capture reference materials, as well as training materials (see Chapter 11 for training materials).

In planning fieldwork and making field preparations, the types of field resources involved tend to vary with the objectives and accuracy needs of a project, whereas the amount of field resources will depend largely on the magnitude and intensity of the fieldwork. Implementing the data collection protocols and coordinating the field resources in order to maximize both data quality and fieldwork efficiency require practice of and adherence to data quality measures and data standards in addition to sound logistical planning. The following chapter addresses the third and final related component of fieldwork planning and preparations, data quality and logistics.

10
Fieldwork Planning and Preparations: Data Quality and Logistics

Introduction

The collection of the highest quality data in the most efficient manner should be paramount to any researcher. While there seem to be countless obstacles to realizing this credo, there are a handful of fundamental data quality measures and logistical concerns that are within the control of the researcher and, when applied correctly, can overcome many more obstacles and place researchers in a much more ideal situation prior to, during, and following fieldwork. Chapter 10 highlights many of these data quality measures and logistical considerations that comprise the third and final component in fieldwork planning and preparations.

Data Quality

Data are only as good as the techniques used to gather, process, and document them. Spatial data are no exception. Moreover, data quality is most at risk during the data collection and data entry phases of a project (Congalton and Green, 1999). For these reasons, data quality should be of utmost importance to the researcher. Besides developing sound protocols and choosing the appropriate equipment for a particular project with certain accuracy needs, there are a number of other procedures the researcher and research team can perform to help ensure high-quality data. Field testing of methods, consistent data collection, data quality control, and data documentation are four such procedures, and depending on the procedure, it might be applied before, during, and/or after the data collection effort.

Field testing

For exceptionally large or complicated projects, such as those involving multiple surveys over large areas using multiple data collection teams or over smaller areas but with intensive, complex data needs, it may be necessary to conduct a field test, sometimes referred to as a pre-test. A field test is performed in order to, among other things, evaluate the data collection methodology, field instruments, and training methods; test the equipment and identify additional equipment needs; identify field personnel issues and needs; and develop a logistical support network. Smaller and simpler projects or follow up visits using previously established methods may require no field testing other than checking the equipment and data forms before leaving the office and training new personnel. It is ultimately the researchers' decision as to whether a field test is necessary, or possible, given budget constraints, but a field test of any sort and for any size project is highly recommended.

Although all projects may not require or be able to afford one, a pilot project is a good example of a field test. In a pilot project, the data capture protocols, database design, data forms, field equipment, personnel, and logistics can be evaluated in a subset of the sample frame. For example, a researcher who wishes to do community mapping for hundreds of villages may choose a small subset of villages for field testing. Rather than a random sample of villages, the researcher might benefit more by intentionally selecting a group of villages with varying field conditions and village characteristics. By doing so, the flexibility of the methodology can be tested and modifications can be made to account for a wider range of situations. While there is no way to fully prepare for all situations that might be encountered, a field test will inform the researcher in ways not previously known and result in a more efficient data collection process during the actual survey work.

Pilot projects have additional benefits. Many times they result in usable data and may actually decrease the number of sites that will need to be visited in the actual survey, provided the methodology does not change appreciably between the field test and the actual data collection. Pilot projects might also be a springboard for writing and receiving grants. With a better understanding of field conditions and data collection needs, a researcher will be able to extrapolate from the pilot study sample and write a more realistic proposal with a budget that is less likely to disappear before the work is completed. Also, a proven methodology may be looked upon more favorably by a granting agency.

A pilot project carried out in the study area may also lead to the identification of local area experts and data collection team members that are willing to participate in the pilot project and, more importantly, the actual survey. By having these individuals participate in both the pilot project and the actual survey, they will likely have suggestions for improving the data collection, and training for the real survey might be a brief refresher for them rather than requiring a full training experience. In this way, the methodology is improved by taking into account local knowledge, and the actual survey may be able to hit the ground running. A pilot project in the study area will also familiarize the researchers and other personnel with the area, if they are not already familiar with it, and give the researchers the opportunity to work on logistics for the actual survey.

Field testing, whether it be a pilot project or simply checking and testing equipment prior to leaving for the field, is a prudent investment for projects of any size. At the very least, confidence in the operation of the equipment is increased. At most, every aspect of the field data collection process can be evaluated, modified if necessary, re-evaluated and streamlined for the actual data collection.

Consistent methods

Once the researcher has evaluated and is satisfied with the data collection methodology, consistent application of the methodology is required throughout the fieldwork as another assurance of data quality. Congalton and Green (1999), in their discussion of collecting reference data for remote sensing projects, point to two areas where researchers can influence the level of consistency in data collection – personnel training and objective data collection methods – which can be discussed in the context of any project collecting spatial data.

Researchers and field personnel should be gathered before data collection begins for an introduction to the project and to one another, and to discuss the overall data collection methodology, the use of GPS technology, and each of their jobs during the fieldwork. This helps each person understand how they fit in with the rest of the project and gives the personnel a sense of purpose and cohesion. Groups of people who will be performing similar tasks, such as a data collection team, should be trained together in training sessions dedicated to those specific tasks, such as the GPS data capture protocols, field forms, questionnaires and possible responses, and equipment. Likewise, personnel with more specialized roles, such as field technicians, team leaders, and local area experts, should be trained individually or in small groups and with their specific tasks in mind. All training should take place just prior to or leading up to the beginning of the fieldwork, or as an initial component of the fieldwork. The methodology is then fresh in the minds of the personnel when the data collection commences. Specifics regarding the training process are addressed in Chapter 11.

Objective data collection is also helpful in ensuring consistency in both the fieldwork and in the collected data. Greater objectivity is possible using more quantitative measurements at field sites as opposed to subjective estimates, but many projects rely on subjective inquiries. At the very least, the GPS measurements themselves are objective, quantitative measurements. The attribute data recorded with the spatial data can be objective or subjective in nature, depending on the questions posed, the observations made, and the methods used to obtain the attribute data. Data collection forms and questionnaires can be designed that limit the number of possible responses and lead all data collection personnel through the identical data collection process, thereby providing a data quality check (Congalton and Green, 1999). Common training of the personnel using these types of forms and questionnaires will ensure some level of consistency as well, but variation in human measurements, estimates, and responses must be expected in projects that are quantitative or qualitative, and the real world does not always fit into predefined categories. Researchers should properly document all known inconsistencies in their methods and data.

Data quality controls

Whenever data are collected or manipulated, error is likely to be introduced. Sources of preventable error in spatial data collection include improper use of the GPS or other devices, leading to incorrect measurements; collecting data about the wrong site; recording information incorrectly in a data form; and entering data incorrectly when transferring it from a hardcopy form to a computer file. Other errors, such as those resulting from malfunctions in equipment, while beyond the control of humans, should be recognized when they occur. Proper training and data quality controls applied during the data collection and data entry phases can help prevent some of these errors and minimize the occurrence of others. Members of the research team who will be collecting and entering data should be trained in the quality controls relevant to their tasks.

Validating the locations obtained by GPS receivers can help identify problems with navigation and site selection, GPS performance, GPS operation, and data recording. Using digital or analog reference sources (e.g. topographic maps, aerial photographs, or satellite images), potential field sites can be carefully identified and marked prior to entering the field. The data collection team can then compare the coordinates obtained by the GPS receiver with the approximate coordinates of each site marked on the reference map. Some software packages accept direct input from GPS receivers and can display coordinates against other spatial data layers in real-time. This gives a data collection team the option of carrying a laptop or handheld PC with the proper software installed, and using it together with the GPS receiver to locate field sites and verify that they have found the correct site. For this to work, the GPS receiver must be configured to output coordinates in the same coordinate system, datum, and units as the reference map.

As an inexpensive and less accurate alternative, researchers might establish an acceptable range of values for the coordinates corresponding to each site. Data collection teams can then verify that the GPS coordinates they obtain fall within the acceptable range for each site. Gazetteers are another alternative for verifying GPS coordinates. They contain geographic places and their geodetic coordinates, which can be compared against the location information obtained using the GPS receiver.

A graphic analysis of data quality and location validation can be achieved by mapping the data using GIS software. The feature locations are examined visually in relation to one another and any existing spatial data layers, such as roads, rivers, and satellite imagery that are in the same coordinate system and datum as the GPS data. Data collection errors can be identified in features that show up in unexpected locations, at unreasonable distances from the study site and the other collected features, in some other part of the study area, or on the wrong side of a road or river. Each of these errors can then be traced back to a problem with the GPS receiver or its operation, incorrect recording of coordinates on the data form, or incorrect entry of the coordinates into a computer file. Depending on the error source, GPS receiver repair or replacement and/or additional training of personnel may be required.

In order to assess the quality of and test the repeatability of the data collection methods, a sample of target sites can be selected, and the data collection process can be repeated by a second, more experienced individual/team. The results between the first and second data collection are compared and any discrepancies resolved. Modifications to the methodology and additional training of the data collection team may be necessary depending upon the results. Similarly, when GPS receivers are used together with other mapping tools, such as rangefinders and clinometers, in order to map distant features, making repeated measurements to the same location using the mapping tools is useful for verifying that the measurement is being performed correctly and consistently. Wide variation between successive measurements would suggest a problem with the measurement technique. If little variation is found, averaging the results is a good data quality control procedure, particularly as the distance between the data collector and the feature being mapped increases and the accuracy of distance and angle measuring tools decreases.

As an alternative to hardcopy data forms, data dictionaries may be used to record attribute data about a site. They decrease chances of data entry error by establishing predefined parameters and by eliminating the need to transfer handwritten records to a digital computer file. As a backup and supplement to data dictionaries, it is always a good idea to carry field notebooks in case the data entry forms in the dictionary malfunction, or in case some special circumstances or site anomalies require additional notation.

Field data management

Proper data management during fieldwork is an extremely important data quality measure that entails data download and backup, data documentation, and data confidentiality. Data is typically a major monetary investment on which all research analyses and results ride. Periodic downloading and backup procedures ensure the safety of that investment and minimize data loss. In the haste to collect and analyze data, data documentation can be easily overlooked. This can also be a costly mistake that undermines data quality and can limit its use and sharing with others. And while not applicable to all projects, data confidentiality should be addressed, particularly with projects using high accuracy GPS to locate homes, individuals, and certain cultural and natural resources.

Data download and backup

Downloading and backing up data is analogous to data insurance. It is a very simple process, yet one of the most significant tasks, the benefits of which are not realized until something unfortunate occurs, such as the loss, irreparable damage to, or theft of the GPS receivers and/or computers holding project field data. With regular downloads and backups, such events are not necessarily complete disasters. As such, download and backup procedures and a regular field data backup schedule should be devised, formalized, and included as part of the training of one or more GPS field technicians and a backup person. The data download and backup procedures and schedule should be followed religiously in order to provide the highest level of data security.

GPS receiver data should be downloaded to a computer on a daily basis, if possible. A one-to-one correspondence between the downloaded files and the files on the GPS receiver should be verified. Upon verification, the data files on the GPS receiver can be deleted from the receiver's memory, freeing up space for the following day's work. Organize the downloaded data appropriately in the computer file system (e.g. by date, site, team, etc.), and immediately make copies of the downloaded data onto floppies, recordable CDs or DVDs, and/or other types of transportable media. Storage media should be labeled logically. As an added precaution, multiple copies might also be made and stored in separate locations away from the computer. In addition to daily backups of data, a weekly backup of all field data is a good idea. While some may joke that one can never have enough backup copies, too many copies can become cumbersome, and some copies may find a way of disappearing. For particularly sensitive data sets, this can be a source of frustration and unease.

Data documentation

One area of data development and data management that is often neglected, intentionally or not, is data documentation. It is understandable though, as it can be a time-consuming and expensive endeavor. Oftentimes, however, an eagerness to collect data and then jump straight to data analysis forces data documentation to the backburner, where it may be put off until a later date or conveniently forgotten. Immediate documentation is important, not only to inform those users unfamiliar with the data, but also to ensure that those who collected the data have a record of its characteristics and limitations. Without data documentation there is really no way for anyone unfamiliar with the data to assess its quality, reliability, intended use, and other characteristics. This is why it is extremely important to generate information about all aspects of the spatial and attribute data and the data collection process. This type of information, called *metadata*, and the standards for creating it, should be carefully considered when planning field data collection.

Metadata is information about data. Some components of metadata for spatial data might include the source of the data; the date the data were collected and the methods used to collect the data; data format; the projection, datum, units, scale, and accuracy of the data; the processing performed on the data, such as real-time differential correction; the completeness of the data; and any known problems in the data. Longley et al. (2001) cite four main reasons why metadata is beneficial for spatial data:

1   metadata allow searches to be performed on geographic data sets,
2   metadata aid in determining if a data set will be useful for a particular research need. (e.g. Is the data of sufficient spatial scale and quality? What are the limitations and acceptable uses of the data?),
3   metadata may contain instructions for handling data, such as data format, compatible software, and size and location of data sets,
4   metadata may provide useful information on a data set's contents.

While not required, it is highly recommended that metadata development be included as part of the data collection process. The field technicians, or anyone managing the spatial data, should be trained and responsible for properly documenting the GPS and attribute data and any anomalies in the data and data collection process. Some GIS packages, such as ESRI's ArcGIS, have built-in metadata development tools. Stand-alone metadata software packages are available as well, such as Intergraph's SMMS. However metadata is generated, it is important that it is created following established and widely accepted guidelines, or metadata standards. The most widely used standard for metadata is the United States Federal Geographic Data Committee's Content Standards for Digital Geospatial Metadata, or CSDGM (Longley et al., 2001). This standard for describing geographic data sets provides a description of items that should be included in a metadata archive. Details regarding the CSDGM can be found at http://www.fgdc.gov/metadata.contstan.html. An alternative source for standards for geographic information and metadata is the OpenGIS Consortium, which can be found at http://www.opengis.org/techno/specs.htm. A useful reference for understanding spatial data quality issues and creating metadata for spatial data is the text, *Elements of Spatial Data Quality*, by Guptill and Morrison (1995).

Creating spatial metadata using accepted standards is a worthwhile investment for projects of any size, but it is not always immediately possible to generate formal metadata. Therefore, in the absence of formal metadata, and at the very least, it is good practice to create a text file for each spatial data layer as part of data documentation and include as much information about the data and the data collection process for that specific data layer as possible.

Data confidentiality

Data confidentiality (see Chapter 7) may not be a relevant concern for all researchers. However, whenever data are collected, both spatial and aspatial, data confidentiality should at least be an initial planning consideration. If maintaining data confidentiality is deemed necessary, it becomes an important data management issue. Data confidentiality should be of great concern for investigations involving human subjects or sensitive cultural and natural resources. Adding a spatial component and spatial locations to such projects only underscores the need to maintain confidentiality.

Some may argue that the availability of certain spatial data, such as high-resolution satellite imagery and aerial photography, undermines confidentiality from the outset. These types of images do provide a picture of the earth and its resources, but the real value in most spatial field data is the discrete spatial location *and* the attribute information linked to and describing certain characteristics of an entity or phenomenon found at that location. Thus, maintaining data confidentiality protects people's identities and characteristics and helps preserve valuable, natural, and cultural resources. As data confidentiality standards and procedures vary from one institution and situation to the next, we do not attempt to provide specific guidelines for maintaining data confidentiality. It is mentioned in the hopes that researchers will recognize where, when and why it is required and will implement and abide by the necessary standards.

Logistics

The second part of this chapter presents a mixture of logistical issues, including some administrative tasks, that should be considered during fieldwork planning and preparations and that apply prior to, during, and beyond the fieldwork, depending on the issue or task. Many of these items are not specific to projects using GPS in data collection fieldwork, and some may not even apply to certain research projects, however they are important in the authors' experiences and warrant mention here. We have organized the various logistical items into five main areas, which by no means represent an exhaustive list.

Project scheduling

A project timeline is an invaluable resource for any research project, and developing a fieldwork timeline is highly encouraged. A timeline provides a synoptic view of the entire project or of a certain phase or group of phases in the project, depending on the focus. Major benchmarks and the approximate time periods that will be required to attain those benchmarks should be included. Figure 10.1 illustrates a sample timeline detailing the plan-

Cumulative weeks

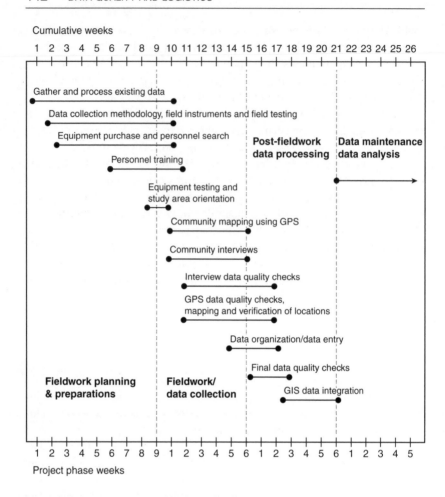

Figure 10.1. Sample project timeline detailing benchmarks in fieldwork planning, fieldwork, and post-fieldwork phases

ning, fieldwork, and post-fieldwork phases of a project using GPS data collection methods for community mapping purposes. The timeline indicates the interrelationships and overlaps among different components of the research and provides a clear, conceptual picture of the research workflow.

Data collection team meetings should be held on a daily basis if possible in order to make announcements to the team; to discuss and attempt to resolve any and all problems encountered, whether they be methodological, technical, safety issues, etc.; to ensure that all data has been downloaded and verified regularly; and to make plans for upcoming work. If possible, weekly meetings should be held between the individuals downloading/managing the data and the spatial information specialist, project manager, and/ or principal researchers. Team leaders and field technicians may provide progress reports at this time and discuss personnel issues and morale problems, equipment problems and needs, problems with the data collection methods or protocols, and similar issues. Daily workplans should be developed by the team leaders and the project manager and implemented by the data collection teams. Time should be allocated in workplan scheduling for data collection teams to return to sites where data was collected incorrectly.

Project housing

Depending on the length and intensity of the fieldwork, a project headquarters may need to be established, whether it is a campsite or a hotel or some other local facility. A headquarters can provide meeting space, a central location for word and data processing, GPS download, and storage space for equipment and other field instruments. If data collection team housing is required, arrangements should be made well in advance for housing with safe and livable conditions.

GPS data collection and government regulations

Depending on the airport, GPS equipment and other electronic devices may raise some eyebrows among the security personnel and baggage screeners, especially where there is increased airport security. For this reason, it is recommended that all GPS equipment and other electronic devices be packed in the checked baggage instead of in carry-on bags, otherwise one may experience delays in the baggage screening area. In terms of packing the equipment, a sturdy trunk is recommended with ample padding material.

Depending on the country in which the researcher is working, the use and/or entry of GPS equipment may be limited or prohibited by the government. Researchers should investigate the governmental regulations, restrictions, and customs fees applied to GPS equipment for the country they will be entering. Hefty customs fees may be demanded for GPS equipment and other electronic devices, such as computers. Permits may be required to use GPS in the entire study area or certain regions of the study area. Check with local, regional, and/or national officials and be sure to obtain the necessary permits.

Accessibility to the study area

Permits and authorizations may be required for the researcher to gain access to the study area or certain regions of the study area. Check with local, regional, and/or national officials and be sure to obtain the necessary permits and/or authorizations. Along the same lines, foreign travel necessitates obtaining a passport and, in some cases, a visa for the dates and amount of time that will be spent in the country. But even with the proper documentation, nature may present its own form of red tape. For example, transportation into and out of a study area may be limited or cut off entirely as a result of weather conditions or road damage due to extreme weather events, such as flooding. Contingency plans should be developed for such occurrences, or it may be necessary to reschedule the fieldwork altogether.

Health and safety precautions

Fieldwork can be both mentally and physically taxing on a research team. Climate, weather, high elevations, the length and intensity of fieldwork, and other factors can exacerbate this fatigue, especially in those who are not acclimated to the environment or the type of work. Members of the research team should understand the study area environment and the conditions under which they will be living and working. Morale among team members can also be affected by field conditions. Principal researchers should be aware and sensitive to these issues and also take some basic precautions:

- Health insurance should be extended to every individual employed by the research project. Health and medical standards can be questionable or nonexistent in some countries and in remote locations. In these situations, health insurance benefits should include trusted international clinics and medical emergency interventions and evacuations.
- Research and data collection team members should be made aware of all disease risks associated with the study area and should receive the proper vaccinations, immunizations, and medications prior to entering the field.
- Adequate first aid kits should be distributed to each member of the data collection team. A more substantial set of medical supplies capable of treating a wider variety of medical problems should be available at project headquarters.
- The safety and supply of food and water should be assessed for the

study area. Trusted sources and an adequate supply of clean food and water should be available to the research team in addition to any backup supplies they carry for themselves.

- Political instability and/or illegal activities in and around the study area put a research team at high risk. If it is absolutely necessary to be in the field under such conditions, ransom and evacuation insurance should be obtained for each researcher in the field.

Summary

As Chapters 8, 9, and 10 have demonstrated, fieldwork planning and preparations are numerous, varied, and can easily take more thought, time, and effort than expected. While there is no way to plan and prepare for all situations that may be encountered during fieldwork, there are many fundamental considerations, decisions, and actions that can be made or taken prior to the fieldwork that ease the transition to the field, increase efficiency and confidence in the data collection effort, and result in high quality data, a true measure of success for any project. We organized fieldwork planning and preparations into three related chapter topics: data and methods, field resources, and data quality and logistics. These topics are arranged in the logical order in which they might be addressed for a research project requiring or benefiting from GPS data collection and spatial information in order to answer the research questions. Depending on the research project, this model for planning and preparing for GPS data collection may or may not be ideal or totally applicable, however, proper planning and preparation are crucial to any project, no matter the size or scope.

11

Transitioning to Fieldwork

Introduction

It is imperative to get data collection right the first time, minimizing costs and crises while maximizing data quality given the uncertainties in all fieldwork. Proper planning and preparations, as outlined in the three previous chapters, can help ensure that the transition to the field and the entire fieldwork phase progress smoothly and result in high quality data. Chapter 11 details the elements of the fieldwork phase from arrival in the field and training through the heart of the fieldwork to the end of fieldwork.

We divide the fieldwork phase into three components and corresponding chapter sections: (1) orientation, (2) data collection, and (3) fieldwork wrap-up. The orientation period includes the arrival in the field of the research team and equipment and the various activities that take place prior to the beginning of the actual data collection, including any training. The second component of fieldwork discusses the actual data collection phase, including the practicalities and troubleshooting of GPS operation, as well as data capture, download, backup, correction, and entry. The third and final fieldwork component, fieldwork wrap-up, includes concluding steps, such as data verification and storage, equipment inventory, and preparing both data and equipment to be returned from the field.

Orientation Period

Depending on the location of the study area, the transition to the field may include a walk outside the building and a drive down the road, or a multilegged flight halfway around the world and a drive of several hundred kilometers to the study area. Some projects may be able to complete their fieldwork in less than a day, while other projects may require months or years to complete the work. Regardless of the length and intensity of fieldwork, there is an orientation period with associated tasks or events that apply to most everyone entering the field for the first time. As one would expect, subsequent trips to the field tend to be less problematic and flow more smoothly due to familiarity with the area and its language, if applicable, and an awareness of potential problems that might be encountered and their solutions. While we cannot put together an exhaustive list of all the tasks and events to expect upon arrival in the field and during the orientation period, there are some tasks that typically demand the attention of the researcher.

Local contacts and initial meetings

If the fieldwork is being conducted in a location a considerable distance from the researcher's home institution, or in restricted access or remote areas, it is important to have established some contacts in the area before work begins. Contacting and meeting with local officials, research collaborators, and/or local area experts are usually two initial tasks to be done upon arrival in the field. Developing a good working relationship with local contacts and including them as part of the research team can be an invaluable resource for projects, as local contacts have greater access to and familiarity with the study area and can make arrangements in the field that otherwise would not be possible or might be very difficult. The use of local contacts will not be suitable for all projects. However, for those projects that do utilize this resource, initial meetings between the researchers and the local research contacts should be held to discuss the proposed research and methods, conditions in and access to the study area, and any special needs and arrangements that should be made prior to and during the fieldwork.

Project headquarters, lodging, and supplies

Local contacts can be instrumental in identifying a project headquarters, lodging facilities, and sources of supplies. Sufficient time and money should be budgeted for identifying, establishing and preparing the headquarters and lodging as well as searching for and obtaining the necessary supplies. For some projects with nearby study areas, the home institution can serve as headquarters, and lodging in the study area may be unnecessary as the area may be reached by car or some other means on a daily basis. Other projects may use a hotel, a local house, or some other facility for project headquarters and team lodging. For those projects with more remote study sites and fieldwork involving overnight or extended stays, an ideal situation is for the local contact or someone else affiliated with the project to find and reserve a location that will serve as the project headquarters and lodging for the research team. However, as this is not always possible, researchers should have contingency plans in place. Sometimes the only option is to set up a temporary campsite for the headquarters and lodging. The project headquarters, lodging quarters, the area surrounding these structures and the study area itself can be ideal settings for data collection training, or training refreshers in cases when the initial training was done prior to traveling to the study area.

Depending on the remoteness of the study area, a large supply of items that need frequent replenishing (food and water, batteries, film, gasoline for generators, etc.) may need to be carried in with the research team or ahead of the research team. When study areas are located within or near cities or towns and transportation is not an issue, packing in large amounts of supplies to the study area and project headquarters is not crucial. Local research contacts should be able to identify trusted and well-stocked sources for supplies.

Contact GPS base station or real-time differential service

If differential correction is to be used as a GPS error correction technique, the source of base station data or real-time corrections should be contacted for final confirmation of service prior to the beginning of the data collection. The researcher should verify that the base station will collect reference data during the same time that the project's rover data will be collected in the field. It is also advised to double-check the settings of the base station GPS receiver with their technician. The provider of real-time corrections should confirm the planned transmission or broadcast of corrections during the time period specified by the researcher. Details regarding the selection of a base station or real-time correction service are found in Chapter 5.

Equipment check

The equipment checklist should be used to account for all field equipment. This same checklist should have been used in packing the equipment for transport into the field. Once all equipment has been accounted for, each piece that requires a power supply for operation should be tested to verify that it is in working order. Manufacturer recommendations should be consulted for the amount of battery consumption to expect, but these numbers vary with the environmental conditions to which the equipment and batteries are exposed. Equipment that does not require battery operation, such as standard compasses, clinometers, and optical rangefinders, while usually quite reliable, should be tested as well.

As part of the field equipment check, it is always a good idea to perform a practice test using the GPS receivers and accessories, any real-time DGPS peripherals, computers, and any GPS software. This usually involves collecting and recording some GPS data and then downloading the data to the computer using the appropriate GPS software and interface cables. This testing should also be included as part of the field tests performed prior to entering the field. By performing a test of the GPS receivers and other equipment and then downloading the GPS data to the computer, the compatibility of the components can be verified and an estimate of the power consumption can be made and used in planning the purchase of DC power supplies or regular access to an AC power supply.

The GPS receivers, computers, and other mapping and field equipment we have discussed in this book are usually quite reliable. However, as with any electronic equipment, there are times when malfunctions occur or when equipment is damaged, lost, stolen, or confiscated. Backup equipment and spare parts are a good idea, but not always feasible, particularly when dealing with receivers and computer equipment that can cost thousands of dollars each. The best advice is to handle and transport all field equipment with extreme care and keep the equipment clean and regularly maintained, if applicable.

Assembling the data collection team

The data collection team may consist of only outside researchers, a mixture of outside researchers and members of the local population, or entirely of locals recruited from in and around the study area. Whatever the case may be, these individuals must be contacted prior to training and fieldwork, and arrangements must be made by them or by the researcher for their transportation to the project headquarters, lodging quarters, training facility, or wherever the data collection team is to be initially assembled. Time and money should be allocated for transportation and for getting the data collection team settled into the lodging facility, if required. Local research contacts can facilitate these types of activities.

Training

Training is a critical component to a successful GPS project. Once the data collection methods and protocols have been developed, then it is necessary to instruct people in how to implement them. However, adequate training involves more than just passing out GPS receivers and field instruments. Members of the field team should have a clear understanding of how GPS works and why it is being used in the project, as well as familiarity with the field instruments. Local area experts should also receive any specialized training.

Training structure

Training in the use of GPS receivers should occur as close as possible to the beginning of fieldwork. Anyone who will be operating a GPS receiver should be trained as well as the local area expert and any supervisory personnel. For smaller projects, GPS training can be completed in half a day and should consist of four distinct parts: (1) Overview of GPS, (2) Receiver-specific Training, (3) Data Collection Protocols, and (4) Practice Session. For larger projects, more time may be needed to cover additional material.

Training materials

Training materials consist of the various lecture material and hands-on train- ing documents used to conduct training of GPS data collection teams and personnel in special support roles. Adequate training of the data collection team and those in special support roles, such as the local area experts and GPS technicians, is necessary to ensure that the data collection methods are applied correctly and consistently throughout the fieldwork. Thus, prepar- ing the proper training materials that cover the relevant topics and provide good hands-on practice for the data collection teams is important. Costs for producing training materials vary with the size and complexity of the data collection effort, but they can generally be measured in terms of the time to construct the materials, including translation into other languages, if appli- cable, and the costs for printing and copying the materials.

Lecture materials
In planning and preparing lecture materials for training the data collection team, the range of topics to be addressed and the lecture format should be developed by collectively considering the research questions and the audi-

ence members' primary language, technical skill levels, and study area knowledge. For example, training a data collection team from the researcher's institution who have worked with the researcher before, have some experience using GPS receivers, and speak the same language as the researcher, can be a vastly different and much easier experience than training a data collection team assembled from the local study area with no technical experience and little or no understanding of the researcher's primary language. Both situations have their pros and cons, but for projects in foreign locales, recruiting a data collection team from the local population can be advantageous for the reasons discussed in Chapter 9. As some projects may not have the luxury of knowing who will be part of the data collection team prior to arriving in the field, preparing lecture materials may be an off-the-cuff exercise for the researcher.

A set of lecture topics is described below which can be applied to a wide range of social science projects. Not all will be appropriate for every project and this is by no means an exhaustive list of topics. They are organized as they might be presented in training a data collection team in the classroom.

1   Project background – this material focuses on the origins, progress, and current status of the research project and its relevance to the features under study, the study area, and the research community in general. Introduction of the principal researchers, research staff and data collection team should be included with a brief background for each person. A brief discussion of the research discipline or the collection of disciplines involved in the project, relevant theories and tenets, and their relationship to the project is useful. However, more time should be spent discussing the features being located and why their location is important to the research. The specific research questions related to the data collection effort should be presented and discussed along with the methodology and any expected results. Personnel roles should be clearly defined and include how they fit into the overall data collection.

2   Study area background – many aspects of the study area should be discussed, including geographic factors (e.g. climate, weather during the data collection, topography, flood regime, etc.), safety factors (e.g. political climate; food and hygiene; diseases; poisonous insect, snakes, plants; dangerous animals; etc.), and cultural awareness factors (e.g. language, religion, taboos, customs, etc.). This discussion may be unnecessary or may need to be modified significantly if the data collection team is comprised of individuals from the local study area population or from somewhere in the vicinity of the study area.

3   Geodesy, mapping, and navigation – since the data collection team will be using GPS receivers, maps, and possibly other mapping tools to navigate and map features on the Earth's surface, some fundamentals of geodesy (projections, coordinate systems, datums, spheroids, geoids, etc.), mapping, and navigation should be included in a lecture or series of lectures. However, due to the complex nature of the topic it should not be too detailed. The primary goals are to make sure the field teams understand the concepts of coordinate systems and datums, that it is important to correctly configure the GPS receiver to the proper coordinate system and datum, and how to correctly record the coordinates.

4   The global positioning system and GIS – this material primarily focuses on the history, design, components, operation, accuracy, error sources, and error correction techniques associated with the global positioning system. Additionally, GPS data and file types and general GPS data collection and integration with GIS and other spatial and aspatial data may be discussed.

5   Equipment checklist and use – this discussion focuses on all the equipment that will be used in the data collection. Equipment assembly, operation, maintenance, and troubleshooting should be discussed as well as all personal items that team members should carry given the conditions of the study area and the intensity and length of the fieldwork.

6   Target feature attributes and field instruments – this material focuses first on the target features under study, including a detailed description of them, their relationships with one another and their surroundings, their behavior, and other properties relevant to the research and research questions. The field instruments should be examined in detail in the second part of this lecture. Occasionally, important aspects of a protocol, survey question, or data form entry, or the entire meaning and intent of these instructions can get altered or lost in a language translation. For this reason, each item and the possible responses should be discussed with the members of the data collection team to ensure clarity and understanding of the data collection methods and field instruments.

7   Spatial and attribute data collection protocols – this material discusses protocols (i.e. guidelines, decision rules, contingency plans) for gathering spatial and attribute data for the target features in the study area. It should describe how to integrate the GPS receivers and other equipment with the field instruments (e.g. data collection forms, field maps and graphics, and reference lists) and use them together for data collection purposes.

8   GPS data processing – discuss the practicalities of GPS data download, differential correction (if necessary), data backup, and management. A detailed discussion of this topic applies more to the field technicians and someone in a data management position. A more general discussion is appropriate for the entire data collection team.

9   Data entry and data quality – a general discussion should focus on all aspects of data entry and methods to ensure data quality, including consistency in methods, quality controls, and quality checks during data collection and data entry, maintaining data integrity during data download and backup, metadata, and data confidentiality. Intense training in these topics should be provided for the team members who will work most closely with the data after it has been collected, such as the team leaders, field technicians, spatial information specialists, and those in data handling, data entry, and data management positions.

Some projects may not have the background, capacity, budget, time, or need to create and present all of these lectures. At the very least, the data collection team should be familiar with the project and research questions, the study area, the target features and their relevant attributes, how to operate the GPS receiver and other equipment, how to record the necessary data using the relevant field instruments, and how to download and backup data and maintain high data quality standards throughout the data collection effort. Projects requiring a more condensed training session might briefly discuss the project background, goals, and study area, and then focus on the basics of GPS technology, GPS operation, data collection, and data processing. Figure 11.1 illustrates a sample contents page for a condensed GPS manual and/or course outline that might be more amenable to projects under time or budget constraints. Providing high-quality training for the data collection personnel will help increase the quality of the GPS data that returns from the field.

Hands-on materials
As a complement to the lectures discussed above, hands-on materials should be developed and provided during GPS training much like a lab component in a university course. Hands-on materials should consist of the following components, and the data collection team should be encouraged to keep the materials for future reference.

1   GPS quick references – the GPS quick reference materials should quickly and easily describe the initialization of the GPS receiver, the configuration parameters and preferred project settings, basics of operating the GPS

TRAINING OUTLINE

I. INTRODUCTION TO GPS TECHNOLOGY
  What Is It?
  The Basic Idea
  Determining Distance Between Receiver and Satellites
  Resolving Timing Errors
  Data Accuracy
  Differential Correction
  Coordinate Systems
  Mapping Datums and Spheroids/Ellipsoids

II. OPERATING THE GPS RECEIVER
  Assembling the GPS Equipment
  Initializing and Configuring the GPS Receiver
  Typical GPS Data Capture Session
  Various Hints/Notes for Operation of the GPS Receiver
  Rover File/Waypoint Log
  Troubleshooting the GPS Receiver

III. GPS FIELDWORK
  Equipment Checklist
  Equipment Assembly and Testing
  Target Feature Attributes
  Navigating to Target Features
  Collecting GPS Data
  Collecting Attribute Data
  Verifying GPS Data
  Troubleshooting Data Collection

IV. GPS DATA PROCESSING
  Data File Transfer Procedures
  Data Backup Procedures
  GPS Reference Station File Acquisition
  Differential Correction of Rover Files
  Exporting Corrected Data to GIS Format
  Loading Corrected Points into GIS

Figure 11.1. Sample outline for a GPS data collection training manual and training course syllabus

receiver, and how to troubleshoot common problems with the receiver. Most GPS receivers ship with quick reference guides of their own, but researchers may wish to customize the reference guides to fit the needs of their fieldwork.

2   How-to sheets – several how-to sheets should be designed covering such topics as a typical GPS data capture session, data downloading procedures, and how to differentially correct GPS data. All steps should be concise and easy to understand, but thorough as well. Including pictures of which buttons to push and the appearance of the GPS and/or computer display at each step is very helpful to the user. How-to materials should be reviewed thoroughly and practiced by the data collection team until all team members are comfortable operating the receivers and downloading the GPS data to a computer. Those members with more specialized roles should understand the differential correction procedures as well.

3   GPS data log forms – a simple GPS data collection log form should be developed and consist of at least three columns: waypoint or rover file name, coordinates, and site description. The data collection team members can use these forms for practicing spatial and attribute data collection at different locations outside of the training facility.

4   Data quality guidelines – a list of applicable data quality guidelines and procedures should be developed and distributed to all data collection and data entry team members. This sheet might be referred to as the "golden rules" of data quality and should be understood and applied throughout the fieldwork and data entry phases of the project.

An important step to include in any training session is an opportunity for hands-on practice. The primary purpose of this step is two-fold. First, it will give members of the field team a chance to become comfortable with the operation of the receivers. Second, it will provide them with the chance to become familiar with the field instruments and data collection protocols. The beginning of the practice session should be unstructured and consist of having the team explore the menus and features of the receiver and familiarize themselves with their operation. Once the entire team is comfortable with the procedure for turning on the GPS units and acquiring coordinates, it is advisable to give them an opportunity to practice the data collection protocols developed for the project. The steps should be as close as possible to what they will encounter once the project is underway. They should be required to complete any field instruments and properly save and name waypoints or rover files. During the practice session trainers and field supervisors can be available to answer questions or solve problems that may arise.

An added benefit of conducting a practice session is that it will serve as a test of the data collection protocols and field instruments. Feedback from the members of the field team can identify aspects of the field instruments or data collection protocols that are confusing, unnecessary, or missing. It is then possible to make slight modifications to accommodate these suggestions. However any changes should be minimal. At this stage it is not advisable to radically alter either the data collection protocols or field instruments unless absolutely necessary. If significant changes are required, it will be necessary to retrain the field team.

Study area familiarity

As part of the training process, the data collection team and other members of the research team should be lectured on various aspects of the study area, including the people and culture, physical geography, weather, safety risks, taboos, customs, and any other important characteristics. Local area experts, if available, should be part of these types of lectures, or at least consulted on the content. When the team has arrived in the study area, a tour of the entire area or selected, accessible or representative parts of the it should be undertaken in order for the research team to familiarize themselves with the layout and to see some of the aspects of the environment that were discussed in the lectures. Sometimes a tour is not possible or practical given the size of the region, or it is unnecessary, as when the data collection team is from the study area. However, it is beneficial for the researchers to get a feel for the locale, in terms of its size, accessibility, ease of travel, and danger zones, prior to the beginning of data collection. Local area experts would be useful at leading tours and/or discussions of the study area using maps and their local knowledge.

An initial tour of the study area might also be an ideal time to identify knowledgeable, local informants. Apart from the local area experts that are recruited by the research team in advance, members of the local population may at times be asked on an *ad hoc* basis to accompany the data collection teams and assist with access to and navigation through certain areas, provide information about certain target features or other elements of the study area based on their local expertise, verify information provided earlier in the data collection by other groups or individuals, facilitate discussion and data collection with other members of the local population, and other needs. Time and money should be allotted for identifying and compensating local informants.

Assigning equipment, territories, and work schedules

Three organizational tasks that should be addressed prior to the start of data collection are the assignment of field equipment, delineation of territories of responsibility and the development of draft work schedules. By this time the research team should have some familiarity with the study area and the data collection personnel should be trained and organized into several teams, unless of course there is only one group or a single person collecting data. The data collection teams should be assigned the relevant field equipment at this time and a log should be kept, which includes a team ID, the names of the members of the team, and a list and description of the field equipment assigned to each team. This same equipment log can be referenced and updated throughout the fieldwork in cases where equipment malfunction, loss or damage necessitates a replacement. At the end of the fieldwork, the log can be used to verify that all equipment is present.

Based on the size of the study area, the distribution of target features across the area, if known, and the number of data collection teams, the study area should be divided into a number of zones, or territories, of responsibility. The delineation of these territories really depends on the nature of the research, the efficiency with which groups of target features can be accessed, and natural, cultural, and political boundaries. For obvious reasons, local area experts, local informants, and data collection team members from the local area should be consulted when delineating these territories. Four basic reasons for defining territories of responsibility are (1) determining a good starting point, (2) planning an efficient way to cover the entire study area in terms of sampling, transportation of teams, etc., (3) developing work schedules, and (4) tracking the progress of the data collection effort. Territories can be numbered and data collection teams assigned to specific areas, and the order in which a study area is surveyed can be determined for a single data collection team or single data collector. Based on the territories of responsibility and estimates for the amount of time required to survey each territory, a draft work schedule can be drawn up consisting of the territory or territories each team will be working in on a particular date. Placing the date and team information on a working map of the study area territories is particularly useful.

An alternative approach to assigning territories throughout the entire study area is to assign smaller, regional territories each day. The first step is to design a plan for the progression of data collection throughout the study area. In this approach, work areas are identified on a daily basis, and multiple teams converge on that area and complete the data collection, then move on the following day. Obviously, the time period of one day is simply for illustrative purposes. This can be done on a two-day basis, three-day basis, or whatever works best for the data collection.

This method is particularly useful if the study area is large and not conveniently centered around the location of the project headquarters and lodging. In these cases, if the teams must move their headquarters and lodging periodically, then it may be best to break up the study area into successive regions. The teams would advance through the study area and focus the data collection in each region.

Data Collection

After the orientation component of fieldwork is complete, the members of the research team are settled in and properly trained, and the field equipment has been tested and is functioning properly, then the actual data collection begins. The first hours, days, or weeks of field data collection are an orientation period in and of themselves, when data capture protocols, equipment, and field staff are put to the initial test and when most problems are likely to arise. This will obviously depend on the complexity, intensity, and length of the data collection. As stated previously, some problems can be dealt with during training, however there are invariably those issues that arise at the beginning of the fieldwork and throughout the fieldwork. Difficulties such as equipment failures, weather conditions, personnel turnover, unforeseen problems with the methodology, and data loss may be encountered and require flexibility and adaptability in methods, time, expenses, and logistics. After the initial kinks have been worked out of the data collection process, the data collection phase enters a period of steady workflow. We discuss the specifics of the data collection in terms of GPS practicalities, data practicalities, and administrative practicalities.

GPS practicalities

In Chapter 8, data capture protocols were outlined for a number of different types of target features and data collection scenarios. Data capture protocols are designed for training the data collection team to adhere to certain guidelines and sets of decision rules. They are also useful for preparing the data collection teams for the unexpected, through the use of contingency plans, while they gather spatial and attribute data for different types of target features and under varying field conditions. Data capture protocols are related more to a research project's objectives, the conditions of the study area and unique target feature settings, and the types of features and attribute data that are necessary in order to answer the research questions.

GPS practicalities, however, include the guidelines and hands-on, step-by-step instructions for successfully operating the GPS receiver as part of the broader data capture protocols. These instructions are more specific to the brand or type of equipment being used rather than the research objectives, research questions, and broader methodology. As equipment manufacturers vary, so do the user interfaces and specific instructions and terminology for operating GPS receivers. We have tried to dilute much of this specificity into descriptions, guidelines, and sets of instructions that can be applied to more than one GPS receiver. Before getting into the step-by-step GPS operation for some typical data capture sessions in the field, there are some fundamental precursors to geographic data capture that are addressed in the next several subsections.

GPS interface

Users of GPS receivers interact with the receivers using a combination of keypad buttons and display menus and/or display pages. Basic, common features of most GPS receiver keypads include a power button, up/down and left/right arrow keys, an enter key, and a previous page/go back button. Depending on the receiver, several additional buttons or a small keyboard may be available. Menus and display pages are typically hierarchical in nature. Several main pages and/or a main menu are at the top of the hierarchy. Subpages and submenus can be selected and accessed from the main pages or main menu.

GPS file types

The coordinates calculated using GPS receivers are typically saved in one of two ways: single position waypoints or multiple position rover files. Most recreational GPS receivers use waypoint files, which consist of a waypoint number and name, latitude and longitude (or the x, y coordinates from an alternative reference system) and occasionally altitude. Mapping grade receivers typically offer both waypoints and multiple position rover files. Multiple position rover files are more robust than waypoints in terms of the positional information they contain (e.g. number of positions, satellite signal properties, position quality, etc.), their potential for differential correction and greater accuracy, and the ability to record accompanying attribute data in the same file using data dictionaries.

Clear view of the sky

As discussed previously, one of the basic requirements of GPS use is that there be a clear view of the sky so that GPS satellite signals are not blocked from reaching the GPS receiver antenna. Clouds and light rain do not appreciably affect the reception of GPS satellite signals, however heavy rain may damage the GPS receiver if it is not weatherproof. Tree canopy cover and nearby obstructions such as buildings, mountains, canyon walls, and even the GPS user's body can interfere or totally block the GPS signal. Using an external antenna or repositioning the GPS receiver may solve some signal blockage problems. However, there are some field conditions for which GPS receivers are not suited and traditional surveying techniques should be employed. The potential for signal blockage or interference is addressed earlier in fieldwork planning. It is reiterated here, as it is fundamentally important to successful GPS data collection.

Antenna stability

The GPS receiver antenna should not be moved while the receiver is calculating and recording coordinates for a location. Doing so can negatively affect the accuracy of the derived coordinates and depending on the surroundings, the GPS receiver may lose the signal from one or more satellites. The only exception to this advice is when the GPS receiver is being used for navigational purposes or is being used in a tracking or route routine to map a linear feature or polygon edge/perimeter. An example might be mounting the receiver antenna on the roof of a vehicle in order to collect data while driving along a road.

GPS receiver initialization

All types and brands of GPS receivers must be initialized. Initialization is usually necessary when the GPS receiver is being used for the first time, after GPS memory loss and rebooting, and when the receiver has been moved substantial distances (800+ km) with the power turned off (Garmin Corporation, 1998). Most receivers can perform the initialization automatically by being switched on and allowed to receive satellite signals for a few minutes. During initialization the receiver acquires satellite orbit, clock, and almanac information and is able to update its internal clock and determine where it is on the planet. In some instances, the GPS receiver may need to be manually initialized by selecting the country or region where the GPS receiver is currently located. Consult the GPS receiver manual for specific instructions.

Basic GPS receiver configuration

Most GPS receivers have a number of configuration or setup parameters that can be altered by the user depending on the receiver and the application. GPS receivers usually ship with the configuration/setup parameters set to some default values. For all receivers, there are several common parameters that should be checked prior to navigation and data collection. Once these parameters are set in the field, they usually do not have to be changed again unless the GPS receiver experiences memory loss, power loss for extended periods, rebooting or re-initialization. They include the coordinate system, datum, units, date and time, battery usage indicator, and waypoint name or rover file name (or prefix).

1   Coordinate system – most GPS receivers have several different coordinate systems in which the coordinates can be displayed and this change can be made within the configuration/setup menu of most receivers. The change is only reflected, however, in the way the GPS coordinates are *displayed* on the screen. Internally, the coordinates will still be calculated the same. Changing the display coordinate system means that the user will see coordinates in the specified system. When using a reference or field map and GPS together to navigate, it is important to change the coordinate system on the GPS receiver display to match the coordinate system used on the map.

2   Datum – the datum parameter can also be changed to a user-defined datum, typically the same datum as any field and reference maps being used to navigate in the field. Most receivers have a large number of datums from which to choose. This parameter is also altered within the GPS configuration/setup menu of most receivers. Just as with the coordinate system, changing the datum only affects the GPS display coordinates. All recorded GPS data uses the WGS84 datum.

3   Units – the units parameter affects the units that coordinates and other data, such as height above ellipsoid (HAE), velocity, and distance, are displayed on the GPS receiver screen. Most receivers offer some combination of metric, English, statute, nautical and custom options for this parameter. The units can be altered within the GPS configuration/setup menu of most receivers and should be set to match the coordinate system units of any field and reference maps being used for navigation.

4   Date and time – the internal calendar and clock of a GPS receiver is usually updated to Greenwich Mean Time or a local study area date and

time after the receiver has been initialized properly. However, the GPS user may wish to display the local date and time, if not already displayed, or some other date and time on the GPS screen. The date and time can be changed within the GPS configuration/setup pages of most receivers by specifying a time zone name or a specific offset from GMT, or by setting the calendar and clock manually.

5   Battery usage indicator – GPS receivers have a battery usage meter that typically displays either the amount of time a set of batteries have been used or a power bar indicating approximately how much battery power remains. Some receivers require the user to reset the battery usage indicator when a new set of batteries has been installed in the receiver. The reset is usually accessed within the GPS configuration/setup menu. If the battery usage indicator is not reset when new batteries are installed, it is difficult to determine how much usage time remains for the recently installed batteries. If a battery usage meter is calibrated for alkaline batteries, the use of rechargeable batteries may cause the bar to report inaccurately, making it difficult to determine the approximate amount of power remaining. It is a good idea to always carry spare batteries when using GPS receivers in the field. Since different receivers may require different sizes, types, and numbers of batteries, consult the GPS receiver manual for information on battery usage.

6   Rover file names / waypoint names – if multiple data collection teams will be working simultaneously in the study area, it is important that there be some way to distinguish the rover file names/waypoint names used by one team from those used by another team. Some GPS receivers incorporate the time and date into the filename and use a default prefix letter at the beginning of the name. Other GPS receivers assign sequential numbers to the recorded GPS data. In both cases, there is the chance that different teams will collect and save GPS data to files with identical names. As an example, to solve this problem, GPS receivers could be configured to use different prefix letters. Assigning each data collection team a different prefix letter (e.g. A, B, C, D, etc.) and making sure that each of their GPS receivers are configured with this unique letter will help avoid potential file conflicts and file overwriting when rover files and waypoint files are downloaded to the computer. In the latter case, a unique prefix letter can also be assigned to each team. Each team will be responsible for saving rover files/waypoints using the naming conventions established by the research project.

Advanced GPS configuration

Depending on the brand and grade of GPS receiver used in data collection, more advanced GPS configurations and associated parameters are available to the researcher, which allow for greater control, flexibility, functionality, and application. We do not attempt to provide an exhaustive list of these advanced configuration parameters. As an example. an application is presented below, which requires advanced configuration by the GPS user, followed by a description of some selected configuration parameters that should be considered in advanced applications of GPS receivers.

Real-time differential correction

Although some GPS receivers may be capable of receiving and processing real-time corrections, they are typically not set up by default to receive those transmissions. As a result, GPS receivers must be configured to accept real-time corrections and these settings should be checked and verified prior to each data collection session. In addition, some GPS receivers require a separate radio to be connected to the receiver via a communication port while other receivers may have a radio built in to the receiver. Basic configuration parameters for real-time differential correction may include activation of real-time mode, selecting a communications port for a separate radio, if necessary, and selecting the transmission/correction format and/or radio frequency. Check the receiver's manual for detailed instructions and parameter configurations.

Common advanced parameters

Some GPS receivers allow the user full control over some or all configuration parameters, while other receivers have factory presets for those parameters that the user cannot change. For those GPS receivers that offer full control over the parameters (mainly mapping and survey grade receivers), it is imperative that users understand the parameters and check them prior to each data collection session. Users are encouraged to consult the specific GPS receiver manual for a comprehensive list and description of available GPS configuration parameters, their default settings and applications. Note that this is not a complete list of all configuration parameters available among all GPS receiver types, but is a short list of some common advanced configuration parameters:

- number of positions – minimum number of positions to collect for waypoint averaging or saving to a rover file;
- logging rates – rate at which positions are recorded for points, lines and areas; rates are often measured in seconds or minutes;

- antenna height – height of the antenna above the ground surface; should be adjusted when using an external antenna;
- signal strength – measure of satellite signal quality; threshold setting; also called signal-to-noise ratio (SNR);
- position dilution of precision (PDOP) – measure of position quality based on satellite geometry; value indicates the maximum PDOP under which the receiver will collect coordinate data;
- elevation mask – minimum elevation above the horizon (in degrees) below which the receiver will not use a satellite to calculate coordinates; and
- 2D/3D mode – two-dimensional location (latitude, longitude) using three satellites vs. three-dimensional location (latitude, longitude, altitude) using at least four satellites.

Typical data capture sessions

This section is dedicated to outlining the basics of operation for some typical data capture sessions using GPS receivers. Manufacturers may differ, but the basic steps are still the same. Refer to the receiver's manual for specific instructions.

Single positions

1   Turn the GPS receiver on.
2   Wait for the receiver to receive signals from enough satellites, depending on the mode (2D vs. 3D).
3   Mark a position. This will be one uncorrected coordinate pair.
4   Save the marked position to memory and/or record data on data collection form.
5   Turn the GPS receiver off.

Averaging multiple positions (waypoint)

1   Turn the GPS receiver on.
2   Wait for the receiver to receive signals from enough satellites.
3   Verify that waypoint averaging is turned on and properly configured.
4   Begin averaging positions over a user-specified period of time or until a user-specified number of positions have been recorded.
5   Save the averaged position or close the file to memory.
6   Verify the coordinates of the waypoint.
7   Record the waypoint name and coordinates on a field instrument.
8   Turn the receiver off.

Rover files with multiple positions

1 Turn the GPS receiver on.
2 Open a new rover file.
3 Begin logging raw coordinates.
4 Pause/Resume as needed.
5 Monitor the number of positions recorded, particularly if the receiver does not have an automatic file closure setting.
6 Close the rover file when complete.
7 Verify the number of positions recorded and the size of the file.
8 Record the rover file name, number of positions, and file size on a field instrument.
9 Turn the GPS receiver off.

Storing positions in a rover file while in motion

1 Turn the GPS receiver on.
2 Set the logging rate to an appropriate time interval to maximize data collection along the route while minimizing the eventual rover file size.
3 Open a new rover file.
4 Begin logging raw coordinates.
5 Traverse the target feature (line or area edge/perimeter); Pause / Resume as needed.
6 Monitor the number of positions recorded.
7 Close the rover file when complete.
8 Verify the number of positions recorded and the size of the file.
9 Record the rover file name, number of positions, and file size on a field instrument.
10 Turn the GPS receiver off.

GPS data download

At the end of each day the field technician should download all rover files or waypoints to a laptop computer or other data-logging device. The number of files downloaded should be cross-checked with the number of collected files reported by the field teams on their field instruments. Any files that do not have a corresponding field instrument should be noted, and the field team that collected the file should be asked to provide further information. Any field instruments that do not have a corresponding file should likewise be noted. If the file is truly missing, it should be recollected later in the data collection.

The exact steps involved in the data download and verification will differ based on the specifics of the project and the data collection effort. However, the established protocols should be followed consistently throughout the entire fieldwork effort. This will ensure the least amount of missing and erroneous data, making for a smoother transition from the field and back to the GIS lab (see Chapter 12).

Data organization and backup

After downloading data to a computer, it should be organized in such a way as to make it easy to identify who collected the data, when it was collected, and possibly the type of features that it represents. This can be done at a variety of levels of detail and is facilitated by good, sound protocols and well-designed field instruments.

In addition to data organization, it is recommended that the field technician periodically back up all data files on to transportable storage media. The media should be stored in a safe and secure location and protected from the weather. These backups are an insurance policy, along with the field instruments, guaranteeing that some form of the GPS data return from the field in the extraordinary event that the GPS receivers are seriously damaged and/or the computer's hard drive is compromised. For added security, it is a good idea, if possible, to occasionally send backed up data to a safe location away from the field site.

Data entry and verification

Depending on the specifics of the project, some data entry and verification can be performed in the field. The advantage to doing it at this stage is that it is possible to catch errors while still in the field and hopefully correct them. The more data entry completed at this stage equates to less data entry during the post-processing phase. However, given the extent of work to be accomplished in the field, the opportunity to do a significant amount of data entry and verification may be limited.

The GPS technician should regularly map the data being collected. Some GPS software includes a mapping tool that will allow the technician to make sure that the data is showing up in roughly the expected locations and not, for example, in a neighboring district or community. If the software being used does not have a mapping tool, it is still possible to verify the data by reviewing the range of coordinates. Using ancillary data such as preexisting hardcopy maps, it is possible to develop a range of acceptable coordinates. The GPS data, once downloaded into a text file or spreadsheet, can be reviewed to make sure the coordinates are within the acceptable range. If it is not possible to define a range of acceptable coordinates, simply reviewing the coordinates that have been downloaded and comparing their values with those previously downloaded is an acceptable approach, although it will only catch the most egregious errors.

If errors are discovered, the next step is to discuss them with the field team that collected the data to determine the cause. If it was a technical problem, replacing their equipment may be necessary, however if it was due to a misunderstanding of the data collection protocols, a review of the training may be in order. Whatever the cause, identifying errors early will prevent them from being duplicated. If possible, erroneous data should be recollected. This can be done by either sending the team out to the site on a subsequent working day, or by recollecting the data in one "clean-up" effort at the end of fieldwork. For an end of fieldwork clean-up effort, one or more teams will be assigned to revisit locations and collect data. This approach also allows the clean-up teams an opportunity to visit missed or overlooked locations.

Administrative practicalities

Administrative practicalities include three daily wrap-up procedures: fieldwork debriefing, equipment inspection and maintenance, and a new work plan for the next day. Since unforeseen problems will arise in the field, a daily fieldwork debriefing is a valuable step. Team members and leaders can discuss problems encountered and potential solutions. This is an ideal time to identify potential barriers to data collection, and adjust the existing protocols to solve the problems.

It is important to inventory and inspect the equipment, performing any necessary maintenance, following each day of work. This can be a time-consuming task, depending on the amount of equipment involved and the number of people performing the inspection. It is a good idea to have the team members inspect the equipment for which they are responsible. This will allow them to continue to gain familiarity with the equipment, while at the same time stress the importance of taking care of the equipment in the field. This will also free up the field technician to perform any necessary maintenance and data-related tasks.

Lastly, before retiring for the day, the teams should meet one last time, and team leaders should plan the following day's events. This may involve the assignment of new territories, reassignment of territories, recollection of bad data, or maintaining the status quo. By performing these administrative tasks each and every day, the whole operation will run more smoothly; obstacles can be surmounted before they can adversely impact the data collection; alternative methods and adjustments to existing protocols can be devised early on in the data collection effort; equipment will last longer, reducing the cost to the project; and the teams will stay focused and interested in their work.

Fieldwork Closure

At the conclusion of the data collection phase of fieldwork, there are several tasks that should be addressed as a final wrap up to the fieldwork. This chapter ends with a discussion of those important concluding steps, such as final data inventory and an inventory of all equipment. Before leaving the fieldwork site, the completed data, supporting materials, and field equipment should be accounted for, organized, and packaged for transport out of the field and back to the office.

Data inventory and storage

The field technician, team leaders, and any other capable personnel should thoroughly check the GPS files against all written records of collected data. If any data are missing, a data recollection effort can be put into action if there is sufficient time. However, once it has been verified that no data are missing, all data files should be saved to transportable media, packed securely, and returned with the researchers to their laboratory for post-field work processing.

Equipment inventory and packing

Similarly, all equipment should be inventoried against the equipment checklist. Any missing equipment should be actively tracked down and located, or recorded as missing. Equipment should be packed for the return from the field in the same manner that they were packed for transport to the field. The equipment should be cleaned before leaving the study area, although it is not always possible to do so.

Summary

This chapter has hopefully shed some light into the many issues involved in

GPS data collection fieldwork. There are many different aspects to a successful field effort, and they all come down to four basic things: proper preparation, sufficient training, flexibility/adaptability, and good personnel.

No matter the location or the size of the project, fieldwork begins with an orientation period where the field teams and the researcher learn about the area, who and what they are working with, and what they are supposed to do. During this stage, any guidance or assistance from local collaborators can be very helpful. Training is another key component of the orientation period. All personnel involved should understand how GPS works, why it is being used and the data collection protocols being employed.

After conducting training and checking the equipment, the fieldwork may begin. While unforeseen problems may arise in the early stages, once the data collection gets underway in earnest, it should run smoothly. Regular verification of data will allow researchers to identify problems while the teams are still in the field. Upon the completion of fieldwork, it is time to return from the field with data securely in hand and begin processing it in order to prepare it for analysis. Post-fieldwork processing of GPS data is the focus of the following chapter.

12

Post-Fieldwork Processing

Once the data has been collected using the GPS receiver, there are some steps that must be taken before incorporating it into a GIS. This type of work is known as post-fieldwork processing. This chapter will describe the steps necessary for post-fieldwork processing of raw and averaged data, as well as differential correction of GPS rover files.

Data cleaning and verification

Regardless of whether data is to be differentially corrected or not, an important step during post-fieldwork processing is data cleaning and verification. If proper data collection protocols have been followed, the amount of data cleaning necessary should be minimal, and will depend on the ultimate destination of your data. This chapter largely assumes that your data will be stored electronically and ultimately bound for a mapping program or GIS. While the specifics may assume a digital destination for the data, the fundamental concepts will apply to analog data.

The first step is to review the collected data. If data has been regularly downloaded from the unit, it is advisable to review the files and make sure that they all are accounted for and all data that were supposed to have been collected were indeed collected. This is best done in conjunction with any field instruments that were used to record information about the GPS data collected in the field. If data are missing then it may still be possible to revisit the site and collect the missing data. If that is not possible, then the data should be recorded as missing. There are several things to look for during this data review process. First, make certain that there are no duplicate file names. If two separate collection teams used the same file naming convention, then it is quite possible that there are duplicate file names for different coordinate files. Another possibility is that the separate teams, if creating their own file names for GPS data files, may inadvertently use the same file name for different locations or accidentally enter the wrong file name. By comparing each collection team's data files with their field instruments, it is possible to catch these errors early and fix them before proceeding too far into analysis.

The second step in the data review stage is to verify that all of the data have been downloaded or recorded in the project's base coordinate system. Different receivers and downloading software offer different options for downloading the data, so depending on the specifics of the program being used, there may be little or no options about the coordinate system. If the coordinates are being entered manually, rather than electronically downloaded to a computer, then follow the same data accuracy protocols that are standard for all data entry procedures. One option involves having one person's data entry verified by a second person following the data entry. The second person compares the hardcopy data entry to the computer file entry. A second option involves two persons sitting at the same computer while entering data from each form. One person can read the data form and enter the information into the computer file while a second person compares the information in the same data form to what is typed into the computer file. This way there is simultaneous independent verification of the data entry. A third option involves having two individuals independently enter data from the same hardcopy data form. Two computer data files are generated and discrepancies between the two files can be easily identified and corrections can be made to the database.

Third, ensure that all data files are in the proper file format for use in any necessary software packages. If post-processing differential correction needs to be performed, then the files should be in the format that allows for this process to take place in the differential correction software. Any other file format-specific needs should be addressed at this stage.

Fourth, make certain that the coordinates have appropriate number formatting. For instance, if the data are in latitude/longitude or another coordinate system in which negative values are used, then it is important to make sure that there is some indication of whether they are positive or negative. This is important regardless of the location of the study area. It is also important to check the number of significant digits in the final coordinates associated with each location. For data files containing one raw coordinate pair or a point-averaged coordinate pair, then the significant digits can be checked immediately. However, for rover files that are to be differentially corrected in the lab, the corrected data contained within the rover file must first be averaged to obtain one coordinate pair that represents the location. Once the averaged corrected rover file is in hand, the number of significant digits can be checked. It is important that an adequate number of significant digits be maintained in the database. Although a common belief is that more significant digits result in more accurate data, the reality is that the number of significant digits should be solely based on the accuracy requirements of the project and the project's coordinate system units. If geodetic coordinates are being used, such as decimal degrees (e.g. 24.428N, 133.849E) in which one degree is equivalent to many kilometers, then obviously a large number of significant digits will provide a high amount of accuracy. There are multiple formats for displaying latitude/longitude coordinates (see Box 4.2). Keep in mind that most GIS and mapping packages require latitude/longitude coordinates to be in decimal degrees (as mentioned above), and it is probably the most commonly used format with GPS coordinates. While it is of course important to get as many significant digits as possible for this format, recording the value to six decimal places for latitude and longitude is a common practice. However, with a coordinate system such as UTM which uses meters as its units, the only reason to maintain coordinates at a finer precision than the meter level (e.g. 243,856 vs. 243,856.012) is for projects with submeter accuracy requirements.

Data verification is another important step and once again if proper collection protocols have been followed, many problems in this area can be caught before this post-fieldwork stage. Simply put, data verification means ensuring that the data are within an acceptable range of values and fall within the expected bounds of the study area. If the spatial information specialist, field technician, or whoever is operating in this capacity, has been checking the data as it has come in from the field, there should be no surprises. However, it is important to review the range of data and make sure that it is within expected values.

Processing the GPS Files

There are several steps required for processing the GPS data after the fieldwork and data verification steps have been completed. In some cases, it is necessary to apply error correction methods to the GPS files. In other cases, it is necessary to average multiple coordinates in a rover file to one location. Data entry and file format management are also common tasks. These steps will be discussed in conjunction with each of the error correction methods (see Chapter 5).

Post-processing differential correction

Post-processing differential correction in most cases requires the greatest amount of post-fieldwork data processing. To review, it is the process of taking coordinates at a known location, referred to as a base station (see Chapter 5 for a discussion on base stations), and determining the offset between calculated coordinates and the known true coordinates. This same offset can then be applied to data collected with a rover receiver at other locations where the true value is not known, with the result being greatly improved accuracy, as long as the rover receiver is within 480 km (300 mi) of the base station and data are collected simultaneously at both locations. Once work in the field has been completed, post-processing differential correction requires that the rover files be corrected using the x and y offsets (known as ΔX and ΔY) and the timestamps in the base station files to modify the corresponding rover file positions.

Processing rover files can be a simple matter of file organization and management. If the GPS receiver is designed to do differential correction, it should internally handle all of the details of formatting and structuring of the files. As noted previously, not all GPS receivers are capable of storing rover files that will permit differential correction. Typically, only mapping grade receivers have the capability to collect correctable rover files, whereas most recreational receivers can only store raw or point-averaged waypoints that contain only one coordinate pair. However, more recreational GPS receivers are being developed that support real-time differential correction, such as WAAS or subscription services. It is advisable to consult the manual for more information as to whether differential correction can be performed on the data collected by your GPS.

Since post-processing DGPS software may have specific requirements about where and how the rover files are stored, it is important to follow the instructions provided by the software. However, in general it is recommended to store the rover files in a manner that allows for easy identification of the date and time the files were collected. This information should be embedded in the file name, but a well-designed directory structure on your computer will facilitate DGPS.

With regard to the base station files, some processing may be involved. The first step is acquiring the base station files. Before fieldwork, when the base station was identified, the means of acquiring the base station files should have been determined at that time. For base stations that constantly collect correction files and post them on the internet, the files can be downloaded following the completion of the fieldwork. For those stations that do not post their files on the Internet, some constantly collect correction files while others have to be contracted to perform this service. For those stations that constantly collect and store correction files, plans for acquiring the files can be done before or after the fieldwork session. However, for the contract-based stations, plans for payment and file acquisition must be handled prior to the start of the fieldwork session. With these last two cases, it can be arranged so that the files are stored on some transportable media, and the files can be picked up in person from the station or shipped.

The last compatibility issue with DGPS concerns file formats. There are multiple formats in which base station files can be stored, some of which are proprietary to a specific company's GPS units. For instance, SSF files are a proprietary file format used by Trimble GPS receivers. Other GPS receivers and differential correction software may not be able to read SSF files. To get around this issue, a standardized universal file format has been developed, known as RINEX files. RINEX (Receiver Independent Exchange Format) was developed by the United States government, and most base station files available from US government-operated base stations are stored in RINEX format. Most of the newer differential correction software should be able to handle RINEX formatted files, however older software may require the RINEX files be converted into another format that the software can read. Consult your software's manual for information concerning its compatibility with RINEX files, and how to convert to another format if necessary.

Because each DGPS software package has its own steps for post-processing differential correction, it is beyond the scope of this book to detail what specific steps are required to implement it. However, in general most software programs will ask you to identify the base station file and the rover file, and will create a differentially corrected file which will need to be saved onto the computer under a different file name that will allow for the uncorrected and corrected files to be distinguished from one another.

After differential correction, each of the rover files still contains many individual coordinate points, although each is much more accurate than before. In order to fully utilize the differentially corrected GPS data, each of these rover files must be modified so that the end result is only one coordinate pair. The most common approach is simple averaging, which can be done in most differential correction software packages. The software averages the x and y coordinates in each of the positions in the rover file and outputs one mean location. As with the corrected vs. uncorrected files discussion above, the corrected mean rover files should be given a unique name to make them easily distinguishable from the other files. This final step provides a file that can now be used in a GIS or mapping software package. Alternatively, some receivers and software packages automatically generate an average coordinate from the positions within each rover file, making this option unnecessary.

Real-time differential correction

Real-time differential correction requires much less processing than does its post-processing relative. In fact, since the error correction is performed on-the-fly in the field, the resultant files only require some minor modifications to make them useful. If the real-time differential correction is applied to a waypoint being collected by a real-time compatible recreational receiver, then the resultant files will contain only one coordinate pair. In these cases, the data are ready to be exported to a GIS-friendly file format. However, if the real-time correction is used with a mapping grade receiver and a rover file, then the averaging technique, described above, for obtaining only one coordinate pair for the data file, must be performed before the data can be used.

Raw or point-averaged files

Processing raw or averaged files is quite a bit easier than the methods already described. File management is a key component, but there could be additional formatting issues that need to be addressed, depending on the way the data are transferred from the GPS receiver.

Typically, once the data are downloaded from the receiver (if the receiver has this capability), the coordinate files can be opened in a text editor or spreadsheet program such as Excel. Problems in formatting can be easily identified and fixed at this stage to make it easier to incorporate the data into a GIS. If the receiver does not have a download capability, the coordinates will have to be entered by hand into a spreadsheet program and formatted, using either the GPS receiver files or the field instruments as the source of the coordinate information. Since it is impossible to provide a universal rule as to what the finished, formatted file should look like, it is recommended that users refer to their GIS for specific instructions on how coordinates should be formatted.

Turning GPS Data into GIS Data

If the steps discussed above have been followed, then the GPS data should be in one of two formats. If the data collected were raw or waypoint files, then the coordinates should have been exported to a text or spreadsheet file, cleaned up, and be organized within a spreadsheet or dBASE file in the format required by the project and the GIS software. If the data collected were rover files, then the researcher should have one or more directories on the computer containing one file for each of the differentially corrected and averaged files. Each of these files should contain one position with one coordinate pair.

There are a wide variety of ways to take files in either of the two formats and turn them into GIS data layers, most of which are software specific. Since it is impossible to cover all of these procedures, two examples will be used that will hopefully provide enough insight into the methodology so that it can be extrapolated to other software and situations.

Spreadsheet/dBASE to ArcView 3.x shapefile

The first example covers the conversion of coordinates in a dBASE or spreadsheet format to a shapefile (.shp) format, which is the proprietary file format for ESRI ArcView GIS software. The first step is to ensure that the coordinate file is structured so that the first column contains the unique identification codes, or primary key. The second and third columns should contain the *x*/Easting/Longitude and *y*/Northing/Latitude coordinates, respectively. Any additional data fields can follow in the remaining columns. It is important to make certain that the coordinate columns are formatted as numeric with the appropriate number of decimal places prior to saving the file. The file should be saved in one of the three formats that ArcView can read: dBASE (.dbf), ArcInfo INFO, or a delimited text (.txt) file.

After starting ArcView, import the coordinate file by clicking the *Table* icon in the Project window and clicking the *Add* button in the same window. Navigate to the directory with the coordinate file, select it, and click OK.

Under the *View* menu, select *Add Event Theme*. In the new window, select the coordinate file and the fields that contain the *x* and *y* coordinates. After clicking OK, the coordinate file will appear in the View window with each coordinate pair represented by a point. To finalize the conversion of the coordinates to a GIS layer, select *Convert to Shapefile* under the *Theme* menu, select a directory, and name the shapefile. The conversion is now complete and the file is ready for all mapping and analytical needs.

Trimble rover file export

This example is different from the last in that the software specific instructions are for the differential correction software rather than a GIS. The software is Trimble's Pathfinder Office, and the file types are Trimble rover files. First, locate all of the corrected, averaged rover files that are to be placed in one GIS layer, and open them in the viewer window. Click the *Export* button to open the Export dialog window. Within this window, numerous options for exporting the data are displayed. Some options can be altered here. Clicking on the *Properties* button opens up a set of tab windows, each of which corresponds to specific export options. The first option is the file format. Some of the possible formats include ArcInfo coverages, ArcView shapefiles, MapInfo MIF files, AutoCAD DXF files, and MicroStation DGN files.

The second option is the coordinate system and the datum. Pathfinder Office contains a wide variety of supported coordinate systems and horizontal and vertical datums. During the export, it will transform the latitude/longitude coordinates to whichever reference system is chosen.

A third option is the type of feature that is to be exported. The choices include *one point per GPS position*, which will output a file with one point per each calculated position in the rover file; *one point per input file (mean position)*, which may be unnecessary (see *Post-processing differential correction* above); *one line per input file*, which will take rover files collected along linear features and connect the points to form a line; and *one area per input file*, which will take rover files collected along the perimeter of an area and connect the points to form a polygon.

A fourth option is the number of attributes that are to be exported along with the features. Some of the exportable attributes are elevation, or height above ellipsoid (HAE); data file name, time-stamp; PDOP and signal-to-noise ratio (SNR); and any data dictionary attributes stored with the rover files. After configuring these options and any others deemed necessary, the software will create a GIS data file with the specified characteristics, which is then ready to be used for mapping and analytical purposes.

Summary

Post-fieldwork processing of GPS data is the last step of the fieldwork phase and prepares the data for the analysis phase. In many ways, this processing of GPS coordinates follows the basic principles of data cleaning and verification that should exist with any type of data collected in the field, regardless of whether it is GPS coordinates or not. It is important, however, to pay close attention to the myriad of factors involved in the processing that are unique to spatial data, and more specifically, GPS data. The type of re-

13

Utilizing GPS Data within Geographic Information Systems

As Chapter 2 suggested, most researchers using GPS are likely to use the results in a spatial context – perhaps to map them, or generate new data by combining the GPS locations with other mapped information, or to analyze the data for spatial patterns. All of these are common functions of geographic information systems. This chapter is a brief introduction to GIS, first emphasizing some basic characteristics that distinguish it from other research tools, and then giving several examples of mapping and analytical functions of use to social scientists. This chapter is by necessity only a brief introduction to these topics, and the techniques and applications covered are by no means a representative sample of the possibilities available to social scientists working with GPS data. It is intended to provide those who are not familiar with GIS some examples of how to integrate GPS and GIS. A list of resources for more information concludes the chapter.

What is a Geographic Information System?

Basic capabilities

A GIS is composed of a set of software tools with capabilities to store, manage, display, and analyze spatial data. Geographic information system software has been steadily growing and expanding in functionality for over 30 years, and it is difficult to give a narrow yet still meaningful definition of what GIS is now, as in addition to a few very widely used packages, there are many other software applications with significant GIS functionality that are designed for use within a particular industry, business sector, or field of research. Nevertheless, we can identify a core set of capabilities required in any GIS for social science.

A GIS must have a sophisticated database management system. Unlike standard relational databases, a GIS database system must be able to store and manipulate the coordinate information inherent in spatial data. Crucially, it must also be designed to maintain a link between each spatial feature in the database (as defined by a coordinate or set of coordinates) and the information about what is *at* the feature's location – its *attributes*. This linkage between coordinates and attributes is fundamental to the value of GIS. It is what enables us to use digital data to represent literally any real-world phenomenon that can be mapped, and then query, analyze, or display that data drawing upon both its "what" and "where" components.

GIS display capabilities generally include traditional data graphing and charting tools, but the emphasis is of course on cartographic display to explore data for spatial patterns or map analysis results. Most GIS software can generate several different common thematic map types, such as choropleth (shaded categorical) maps, dot density maps, and contour plots, and many have considerably more advanced and flexible display capabilities. Furthermore, the limitations of traditional, static, paper maps do not apply to GIS. Since GIS mapping routines operate on digital databases, a display can be "zoomed" in and out to display at any scale, and if one map design or set of symbolization choices is not satisfactory, then the parameters can be changed and the map redisplayed on screen or reprinted.

Tools for analyzing and querying GIS data are capable of functioning on data based upon spatial location, attribute characteristics, or a combination of the two. For example, a purely spatial query might involve selecting all census tracts that border a hazardous waste facility, while an attribute query might instead select all census tracts in any area that have a specific characteristic, such as percent in poverty or racial composition. A powerful capability available only in spatial databases such as a GIS is to perform a combined query or analysis, such as locating all residential areas with populations over 2000 people and within a 15 minute drive of a public school. Such a query actually creates information that did not exist independently in either the coordinate or attribute data, a hallmark of a GIS as compared to aspatial database systems.

There are a number of analytical tools that are common to nearly all GIS applications, such as the query capabilities described above, overlay analyses (e.g., where do public buildings and flood zones overlap?), point pattern analyses (e.g. are cancer incidents in a neighborhood randomly distributed or clustered in a particular area?), and more. Several examples of these are given below. But many GIS software packages are adaptable to almost any spatially focused question. Many provide scripting or programming tools to create custom analyses, and most are capable of exporting and importing data to and from other research tools, such as statistical analysis packages.

GIS databases

One characteristic that is explicit in nearly all GIS software, although the terminology used to describe it may differ, is the use of *map layers* as an organizing principle for spatial data in a GIS. One of the original motivations for the development of geographic information systems was to computerize the tedious process of preparing maps on transparent film and overlaying them, one on top of another, in order to study the combinations of phenomena that appeared on the different map layers (see *Map overlays*, below). This metaphor of map layers has persisted as a basic organizing principle for GIS: each different type of mapped feature (roads, census tracts, household locations, etc.) in a GIS database is stored in a separate layer, with all the layers coregistered in the same coordinate space. Since the GIS is manipulating digital bits instead of analog information, this process of layering is much easier now.

There are two different ways of storing spatial and attribute information and the linkages between them: the *vector* and *raster* GIS data models. Each of these data models has advantages relative to the other, and the more functional GIS packages implement both of them. While the implementation specifics for these two data models differ among GIS software packages, there are some characteristics common among all of them.

The vector data model is implicitly discussed throughout most of this book. It uses points, lines, and polygons to represent phenomena on (or near) the Earth's surface. This is essentially a digital representation of the system used by a traditional cartographer when drafting a paper map – points are used for simple locations, lines are drawn to represent linear phenomena such as roads, and polygons represent area features such as water bodies or political units (Figure 13.1). The differences between a vector GIS database and a traditional map are twofold. First, in a GIS database, the points, lines, and polygons are stored as a single coordinate pair (in the case of a point), or a series of coordinates (for lines and polygons). Second, the characteristics of the phenomena being mapped are stored in a database table as attribute information, instead of being represented with a particular symbol size, shape, or color on a map. Each point, line, and polygon in a

Vector features

Attribute tables

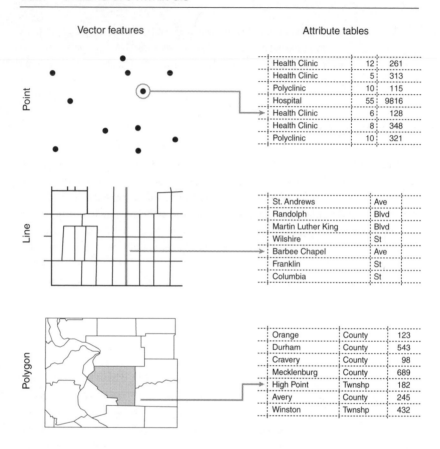

Health Clinic	12	261
Health Clinic	5	313
Polyclinic	10	115
Hospital	55	9816
Health Clinic	6	128
Health Clinic	8	348
Polyclinic	10	321

St. Andrews	Ave
Randolph	Blvd
Martin Luther King	Blvd
Wilshire	St
Barbee Chapel	Ave
Franklin	St
Columbia	St

Orange	County	123
Durham	County	543
Cravery	County	98
Mecklenburg	County	689
High Point	Twnshp	182
Avery	County	245
Winston	Twnshp	432

Figure 13.1. Vector point, line, and polygon features. Each feature has a corresponding record in an attribute database

database has a single, corresponding record in the attribute table. With vector data, the GIS recognizes the point or line or polygon as a distinct entity with a location and attributes.

In the raster GIS data model, a regular array of grid cells is extended across the area being mapped, and characteristics of the phenomena being mapped are recorded at each cell location (Figure 13.2). In this case, the cell values themselves are the attributes. Unlike the vector data model, the raster approach is not analogous to what we think of as a traditional map, but it is the data model used by satellite imagery and aerial photography, both of which are important GIS data sources.

Raster data
e.g. Elevation (m)

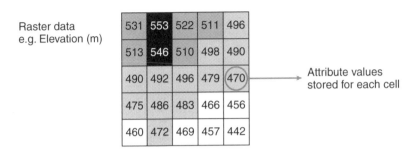

Attribute values
stored for each cell

Figure 13.2. An attribute value is stored for each cell in a raster map layer

In practice, most GIS software packages have the ability to convert data between the raster and vector data models (usually incurring some information loss). This capability is frequently used, sometimes simply to get all the elements in a GIS database into a single standard format. Some GIS analyses also lend themselves better to one data model or another, so portions of a database may be converted solely for the purpose of employing a particular analytical technique.

Making Use of GPS Data in a GIS

Mapping

Data visualization is an important component of any research project, essential to both hypothesis generation and validation. Utilizing GIS mapping tools to display and explore GPS-collected data is a crucial first step in making use of the data. Frequently, simply displaying collected points in mapped form will begin to tell a meaningful story, much as a quick scatterplot can with aspatial data. In fact, a map of GPS point data is itself essentially a scatterplot, merely one that has coordinate eastings and northings as its *x*- and *y*-axes. The map of violent robbery crimes in Figure 13.3(a) illustrates how a simple point map can reveal patterns in the collected data – clearly the incidents are occurring in clusters.

But the real power of GIS comes in its ability to map multiple data layers in a single coordinate space. Note how much more is revealed to a researcher interested in associative or causal relationships when other map

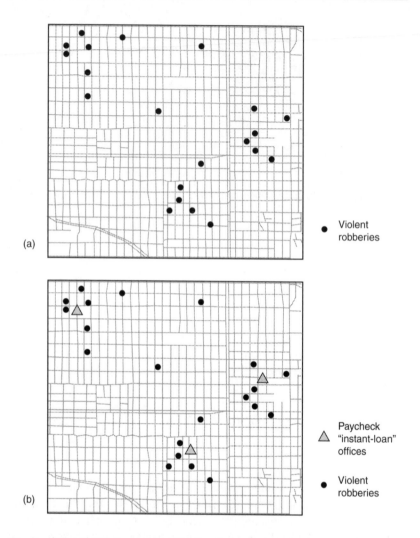

Figure 13.3. (a) Point map showing locations of violent robbery crime events. (b) The same map but with map layers containing locations of paycheck cashing and instant-loan businesses displayed as well

layers are added to the crime locations in Figure 13.3(b). Once a map layer containing the locations of paycheck "instant loan" businesses is displayed with the map of crime locations, it becomes apparent that not only are the violent robberies occurring in clusters, but there is a clear spatial association with the businesses that advance significant amounts of cash to patrons, usually low-income, for the exchange of a forthcoming paycheck.

Map overlays

Social scientists frequently work at the individual, household, or community levels of analysis. One important use of GIS is the development of higher-level, contextual data for study subjects. For example, a researcher could add information about the communities within which they are located to the data for individuals or households. Or if communities are being studied, a researcher might want to know what higher-level political entities the communities fall within, or know something about the natural environment surrounding the communities.

GIS can attach information about the higher levels of aggregation to the objects of study using a process called *map overlay*, which as its name implies, can be visualized as lining up one map layer on top of another and then combining information from the two layers. A common social science example is to associate the socioeconomic or demographic characteristics of a census enumeration unit with a GPS-collected location, as shown in Figure 13.4.

This is among the most basic of GIS operations – what happens in a map overlay is essentially the transfer of attributes from one map layer to another. The GIS determines which polygon in the overlay layer each point in the input layer falls within, and the attributes of the overlay polygon are transferred to the input point in a process known as a point-in-polygon overlay. Note in Figure 13.4 that after the overlay, the attribute record for each household location now contains the census information for the district within which it is located.

One of the benefits of analysis within a GIS environment is that these sorts of cross-layer comparisons and operations can be performed on literally any combination of data, as long as all the layers involved are spatially coregistered within the same coordinate system. For example, many other sorts of contextual data can be generated using a GIS, in addition to the example of census-derived socioeconomic data given above. An anthropologist studying population–environment interrelationships in agricultural communities in a developing country could generate a variety of contextual measures about the physical environment for each community. For the household location above, a point-in-polygon overlay was used, but the layers involved in the overlay can be any sort of vector or raster data. Figure 13.5 illustrates how the anthropologist might generate environmental contextual

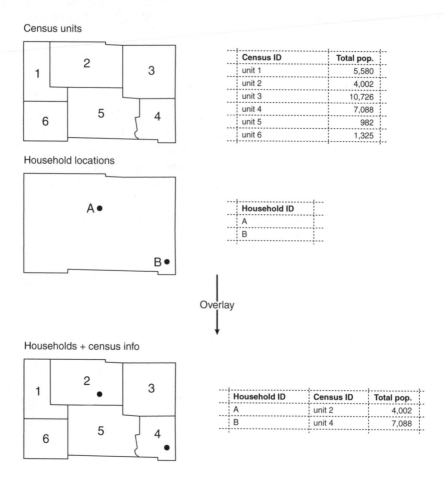

Figure 13.4. A point-in-polygon overlay can attach contextual socio-economic data from census enumeration units to GPS point locations of households

data for study villages using vector polygon soils data and a raster data layer of land cover types derived from satellite imagery.

Simple distance analyses

Since the goal of using GPS for data collection is to find out where things are, it is not surprising that social scientists with GPS locations frequently

Soils (polygon)   Landcover (raster)

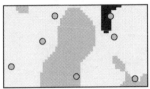

○ Study villages

Figure 13.5. Overlaying polygon and raster map layers to generate information about the environmental conditions surrounding communities

want to ask "What's nearby?" or "How far is it to the nearest _____ ?" Proximity and distance operations are also basic GIS capabilities, and a variety of them can be found in most GIS software packages. Measuring distances is easily accomplished within a GIS. Recall from the discussions in Chapters 4 and 7 that one of important decisions to make for a GPS project is to select an appropriate projected, or planar, coordinate system as the standard coordinate reference base for the project. Assuming the layers in a GIS database (GPS-collected or otherwise) are referenced to a planar coordinate system, then measuring the distance between two points x_1,y_1 and x_2,y_2 is a simple matter of employing the Pythagorean theorem to calculate straight-line, or Euclidean distance:

$$distance = \sqrt{(x_1 - x_2)^2 + (y_1 - y_2)^2} \qquad (13.1)$$

This simple calculation is the basis for a variety of proximity- or distance-based GIS operations.

One common use of distance measurements is to develop information about the neighborhoods surrounding study sites using GIS *buffering* operations. An urban sociologist concerned with the effect of neighborhood environments on children's social and physical development might be interested in the recreational opportunities in the vicinity of children's households. Instead of relying on a basic count of the number of recreational facilities located in administratively defined neighborhoods, the sociologist could use a GIS to build buffers at a specified distance around each household and then count the number of facilities for recreation that fall within each buffer (Figure 13.6). This would generate a measure of neighborhood recreational opportunities specific to each household, and wouldn't rely on

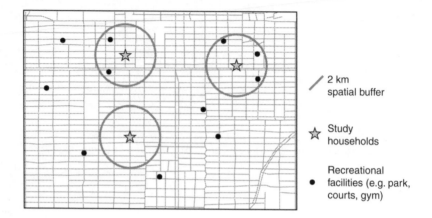

Figure 13.6. Recreational facilities and spatial buffers around urban households

official neighborhood boundaries, which might not reflect the children's true activity spaces.

Another basic distance operation is a *minimum distance* analysis, in which two point data layers are compared to determine, for each point in one layer, the location of the nearest point in the second layer. Consider an economist studying households in a developing country and their access to markets for selling produce they have grown. One item of concern would be the distance from each household to the nearest market. Assuming the specialist's fieldwork included collecting GPS positions for the markets and households within the villages being studied, a minimum distance analysis would determine which market is closest to each household (Figure 13.7), and the distance between it and the household could be used as a measure of access.

Network distance/cost analyses

The economist's distance calculation described above is based on Euclidean distance. That is, it represents a straight-line distance as the crow flies between the households and the market locations. However, in many settings people are much more likely to travel faster, but less spatially direct, routes along local transportation networks, whether on foot or by car, bus or other automated transport. If a data layer is available containing linear features

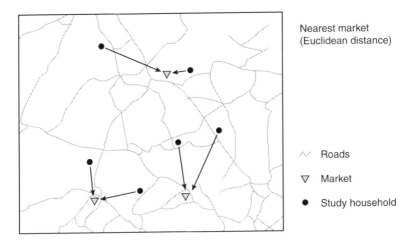

Nearest market
(Euclidean distance)

∿ Roads

▽ Market

● Study household

Figure 13.7. The nearest market location to each household, based on Euclidean distance

representing the local transport system, then distances that more accurately assess the actual distance traveled can be measured using a GIS.

Network analyses model travel or flow networks and calculate distances along linear segments of the networks. In the distance to market example, let's assume that villagers are likely to have access to a moped, motorcycle, or automobile for their travel. The GIS would assign each household and market location to the nearest road segment to determine "starting" and "ending" locations within the network, and then length of the road segments along the shortest path between each household and each market would be summed to determine the actual travel distances between the locations. Then the shortest road distance between each household and the markets would be selected to provide a more accurate assessment of the actual distances traveled by the villagers (Figure 13.8).

By incorporating more attribute data into the analysis, it is possible to refine the travel calculations even further by considering travel *costs*, such as time or fuel expenditure. For example, people making travel and destination decisions are likely to factor in not only the distance involved to reach the necessary destination, but also the time required to travel the distance. If the economist added attributes to the database such as road surface, road width, and speed limit along each road segment, it would allow the calculation of an expected mean travel speed for each segment. Multiplying this travel speed for each road segment by its length

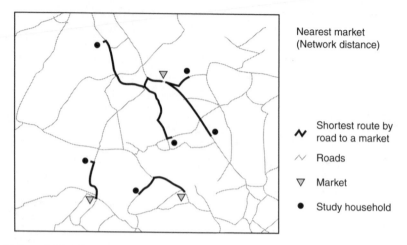

Figure 13.8. The nearest market locations to households, based on network distance

would create a new attribute that estimated the actual cost – in this case, expenditure of time – required to traverse it. Then a network analysis would sum the segment times along routes to the markets, and compare the totals to determine the nearest market location in terms of the travel time required.

Examining point densities and distributions

Mapping data was described above as a very useful tool for exploring the distribution of spatial features across or within an area of interest. Sometimes quantitative assessments of spatial distributions are helpful as well, and they can easily be accomplished within a GIS. Several simple measures of point data distribution are useful examples.

A city public health department with household GPS locations of occurrences of an infectious disease with an unknown source could use a GIS to map and explore the spatial distribution of the outbreaks. Displaying the disease point locations with data layers of the infrastructure, waterways, housing, and other relevant features would provide an initial, visual assessment of the distribution. Calculating the *mean center* of a set of points is the two-dimensional equivalent of calculating an arithmetic mean, and it provides a similarly useful quantitative measure of central tendency. The calculation is simple; it is the

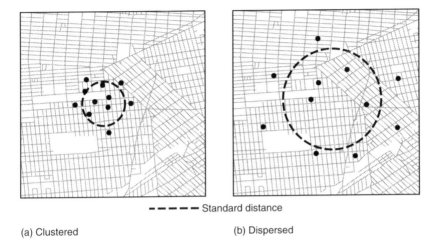

------ Standard distance

(a) Clustered   (b) Dispersed

Figure 13.9. Standard distance plotted on (a) clustered and (b) dispersed point data

sum of the x and y coordinates, each divided by the total number of points: If (X_i, Y_i) is a set of n points, $i = 1, 2, \ldots, n$, then the mean center $(\overline{X}, \overline{Y})$ is calculated by:

$$\overline{X} = \frac{\sum_{i=1}^{n} X_i}{n} \qquad \overline{Y} = \frac{\sum_{i=1}^{n} Y_i}{n} \qquad (13.2)$$

Calculating the mean center of the incidences would summarize the spatial distribution of the disease locations, and if the disease had spread from a single source of infection, the mean center would give a potential starting point to investigate the source. If the date of onset were among the attributes for the disease location point data, then calculating the mean center at a series of time steps would capture the spatial pattern of its movement within the population. Comparing the movement of the disease to social and environmental information in other map layers would provide clues regarding its origin and transmission.

A useful complement to measures of central tendency is an assessment of dispersion. *Standard distance* is the two-dimensional equivalent to a unidimensional standard deviation. It is based on the distances of points from their mean center. The standard distance (S) of a set of n points is given by the following formula:

$$S = \sqrt{S_x^2 + S_y^2} \quad \text{where}$$

(13.3)

$$S_x = \sqrt{\frac{\sum_{i=1}^{n}(X_i - \overline{X})^2}{n}} \quad \text{and} \quad S_y = \sqrt{\frac{\sum_{i=1}^{n}(Y_i - \overline{Y})^2}{n}}$$

A spatial distribution of the disease locations with a smaller standard distance (tightly clustered) instead of a widely dispersed set of points might suggest to the public health officials that the disease was spreading from a single source, not many (Figure 13.9).

Another question of interest to the public health officials is whether or not the disease locations display a regular or a random spatial pattern, or are they in fact clustered? One of many clustering tests is the *nearest-neighbor* statistic, which assesses whether the distances between points and their nearest neighbors is greater than, less than, or equal to what would be expected in a spatially random distribution of points. Determining the nearest-neighbor statistic involves calculating the distances between every pair of points in the data layer, a process too lengthy to consider doing manually for more than a few points, but easily accomplished within a GIS (a single command in many GIS software packages). The standard distance (R) of n points within a bounded area is given by:

$$R = \frac{\overline{r}}{\overline{r}_e} \quad \text{where} \quad \overline{r} = \frac{\sum_{i=1}^{n} r_i}{n} \quad \text{and} \quad \overline{r}_e = \frac{0.5}{\sqrt{n/_{AREA}}}$$

(13.4)

AREA is the size (spatial extent) of the bounded area within which the points are located, r_i is the distance to the nearest other point, making $\overline{r}$ the average nearest-neighbor distance, and $\overline{r}_e$ is the expected average nearest-neighbor distance (in a random distribution). A nearest-neighbor result less than one indicates that the points are clustered more than would be expected randomly, while greater than one indicates a regular distribution—one in which the distances between points are maximized. A result equal to one indicates no clustering (Figure 13.10).

Nearest neighbor statistic

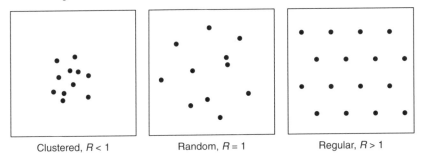

Figure 13.10. Spatially clustered, random, and regular distributions of point data

Summary

A remarkable variety of spatial analyses can be performed with a GIS, given sufficient data and the appropriate software. Some GIS software packages are geared more towards data visualization and mapping, while others emphasize analysis, and there are an increasing number that are directed towards specific fields of inquiry. One characteristic of academic research, including in the social sciences, is that it frequently leads researchers to ask new sorts of questions that may not readily be answerable using existing software tools, even within more advanced and extensive suites of GIS software. Fortunately many available packages incorporate scripting or programming languages that allow custom tools to be created, essentially making the range of analytical possibilities infinite. The examples in this chapter barely touched upon what is possible even with many of the more basic GIS tools. Those interested in a more comprehensive review of the field should consult the list of resources for more information, below.

BOX 13.1

Resources for More Information

Books

Chrisman, N. 2001: *Exploring Geographic Information Systems*, 2nd edn, New York: John Wiley and Sons.
An introductory GIS text with special emphasis on understanding the characteristics and limitations of spatial data and measurement frameworks.

Longley, P.A., Goodchild, M., Maguire, D.J., Rhind, D.W. 2001: *Geographic Information Systems and Science*, New York: John Wiley and Sons.
One of the best introductory texts, with good coverage of both practical and theoretical issues.

Liverman, D., Moran, E. F., Rindfuss, R. R., Stern, P. C. 1998: *People and Pixels: Linking Remote Sensing and Social Science*, Washington D.C.: National Academy Press.
Nominally about the use of remote sensing in the social sciences, this National Research Council publication includes articles from a variety of research projects utilizing GIS and incorporating GPS and remotely sensed data. A good illustration of a variety of possible uses for GIS and GPS within the social sciences.

Software

ArcExplorer: http://www.esri.com/software/arcexplorer
A free GIS data viewer available from Environmental Systems Research Institute (ESRI), one of the most prominent makers of GIS software. A good starting point for beginners interested in exploring GIS capabilities, it has basic data display and query functions.

CrimeStat: http://www.icpsr.umich.edu/NACJD/crimestat.html
CrimeStat, a spatial statistics program free to education and research users, implements a wide variety of point-pattern analyses. Developed for the spatial analysis of crime incidents, its analytical tools are applicable to many social science disciplines. It exports data in formats readable by many GIS packages, including ArcExplorer (above).

EpiMap 2000: http://www.cdc.gov/epiinfo/
EpiMap is a module included in the program EpiInfo distributed by the United States Center for Disease Control. It is public domain software developed for Epidemiology and public health research. EpiMap is a modified version of ArcExplorer that has been integrated with the main program EpiInfo. It can import geographic boundary files that can be used in conjunction with an EpiInfo dataset and produce basic maps.

SIG-EPI: http://www.paho.org/English/SHA/shasig.htm
SIG-EPI is a GIS produced by the Pan American Health Organization (PAHO) to assist health planners and researchers. It imports data in a variety of formats and can perform a variety of epidemiological tasks. While there is a charge for the software it is less than most commercial GIS packages.

14
Conclusion

Writing about incorporating GPS into a research project is in many ways like trying to write a book about how to drive a car. Simply saying, "Turn key, press foot on gas pedal to go, press foot on brake pedal to stop and steer in the direction you want to go" misses many important steps and does not provide guidance about what to do in bad weather, when the car needs fuel and laws the driver must obey. Likewise, there are many details that could vary depending on the type of car, the destination, and the environment around the car (urban vs rural, the weather outside, etc.). However, after reading about how to drive a car, the reader, once behind the wheel, would discover that certain things become obvious while others things discussed in the book are irrelevant to their particular situation. If the book is well written though, there will be some things covered that will save the driver time, make the car run smoother, and maybe even prevent an accident.

This should be kept in mind with regard to GPS. It is impossible to write a one-size-fits-all book about using GPS. Each individual project will have its own needs and the details will vary depending on the specifics of the project. However, once underway, researchers will find that many of the steps discussed in this book will be obvious, while other aspects may not apply to their project. It is the goal of this book to provide instructions and guidance for understanding and using GPS that will save researchers time, money, and even provide inspiration about the roles that GPS or related spatial technologies can play in their projects.

If the contents of this book could be summed up in one word, it would be planning. In order for any fieldwork to be successful, good planning is crucial, and working with GPS is no exception. In fact, every word in this book serves to underscore the importance of considering and planning what needs to happen prior to, during, and following fieldwork in order to achieve a project's research goals and objectives.

Admittedly, planning can be a difficult process, especially for those unfamiliar with GPS. However, the material presented in this book can be used to guide novice GPS users through the process, give an indication of what to expect, and even shed new light on the subject for more experienced GPS users. To that end, it would be beneficial to summarize the material presented in the book.

Identify Research Goals

Any research endeavor requires a clear understanding of the goals and objectives being pursued. Understanding the research question(s) or hypothesis(es) will bring a focus to the fieldwork and provide a framework for the data collection protocols and data processing and analysis.

Determine Accuracy Needs

With the right preparations and the best equipment, GPS can provide subcentimeter accuracy. However, in the hands of an uninformed user, even the best equipment can place something tens or hundreds of meters from its true location. The accuracy needs of most social science research will fall somewhere in between these two extremes. The accuracy requirements of a project are primarily a function of the goals of the project, the features or phenomena being studied, and money. A general rule of thumb is the greater the accuracy being sought, the greater the cost required. Receivers capable of providing the highest positional accuracy usually cost more than receivers that may only provide minimal error correction. The level of detail of other primary and ancillary data also influences accuracy needs. Existing governmental or organizational map accuracy standards may be referenced when determining the accuracy needs for a project.

Select GPS Receiver and Error Correction Method

Once the accuracy needs have been established, it is necessary to use the GPS receivers and error correction methods that will yield the required accuracy. Mapping grade receivers are specifically designed to accommodate higher accuracy data collection. They can provide many specialized features such as data dictionaries, software to help plan optimal times for data collection, and extended memory capacity. These receivers also provide greater control over the conditions under which the receiver operates, allowing the user to eliminate certain satellites and control for PDOP, or the position of satellites in the sky. Recreational GPS receivers are designed for use when hiking, bicycling, or other nonresearch activities, but they can be used successfully in many research projects. Their features typically vary with their cost, and the number of features is usually less than that found in mapping grade receivers.

Also linked to the selection of a GPS receiver is a decision about what error correction method (if any) will be employed, as this decision influences the type of receiver needed for the job. Differential correction provides a high level of accuracy, but requires a receiver capable of collecting data that can be differentially corrected. Post-processing differential correction requires that a base, or reference, station exist somewhere near the fieldwork, while real-time differential correction requires a more specialized receiver and, in some cases, an expensive subscription to satellite-based correction services. Position averaging can provide a moderate level of accuracy and be performed with mapping grade receivers and many models of recreational receivers. Operating with no error correction, or using raw coordinates, is also an option. If no error correction method is used, then the accuracy of the collected coordinates will feel the full effect of errors or variations in the GPS signal.

Determine the Coordinate System and Datum

It is tempting to give short shrift to this aspect of GPS; the concepts can be complicated and difficult to understand and convey to the reader. However, coordinate system and datum misunderstandings can add significant error. Therefore, it is essential to determine the coordinate system and datum being used and to adjust the settings on the GPS receiver to record and display in the established coordinate system and datum.

All GPS receivers convey information about the user's location by reporting coordinates, either in latitude/longitude or one of many possible planar coordinate systems. The coordinate system selected for use in a project should be determined by its commonality with what is used in other project spatial data as well as the analyses to be performed on the data. If there is no clear consensus on which coordinate system to use, latitude/longitude in decimal degrees may be used as an interim solution and the collected data can be projected into a planar coordinate system at a later date.

Just as important is determining the datum to be used. A datum is used to link an idealized shape of the earth (an ellipsoid) and a coordinate system defined on it with physical locations on the actual surface of the Earth. If the datum of the collected coordinates is different from the datum of other spatial data with which it is used, then positional error will result between the data sources. Therefore, as with coordinate systems, it is important to select the project datum based on its commonality with the other spatial data in the project, and verify that the GPS receiver records data in that selected datum. WGS84 is a commonly used GPS datum and would be a safe choice if it is unclear which datum should be used.

Identify Target Features

At this stage in a project, fieldwork planning and preparations begin by determining the specific target features or types of features that will be located and how they will be located. The research goals and objectives should guide the identification of target features and the information to be gathered about the target features as well as ideal ways to represent, or symbolize (e.g. point, line, polygon) features for analysis and modeling within a GIS.

Develop Data Collection Methodology

Once the target features have been identified, a data collection methodology for locating and describing the target features needs to be developed. If a simple point location of a feature or phenomena is needed, then the data capture protocol can be as simple as deciding where the field team member will stand with the GPS receiver while the coordinates are collected. If more complex features, such as lines or polygons, are being located and described, then the data capture protocols will be more involved. As part of this process, contingency plans should be developed to guide field teams with what to do in the event that protocols cannot be followed. The data collection methodology should also include the details necessary to support the selected GPS error correction method, if any is to be used.

Assemble Field Resources

Based on factors such as size of the study area, number and location of target features, time available to complete fieldwork, and the specifics of the data collection methodology, it is necessary to determine and acquire the relevant types and amount of equipment, personnel, and field instruments needed to successfully complete the fieldwork. Besides GPS receivers, the fieldwork may require computers and accessories, additional mapping tools, photographic and communication devices, personal gear, and many other items. It's important to fully understand the utility, operation, strengths, and limitation of the equipment.

A field team might consist of one person or scores of people. As the number of people collecting data grows, the number of support personnel grows as well. A project manager may be needed to oversee budgets, time lines, and other personnel. Field technicians who are familiar with the global positioning system and receivers can troubleshoot technical problems and download data periodically. If working in a foreign country or an unfamiliar location, then a local area expert can serve as a liaison to the community or region and advise about GPS restrictions, study area accessibility, and other potential obstacles in the area.

Field instruments consist of the data collection forms, reference maps, graphics and lists that facilitate the recording and organization of field data. These documents are tailored to meet the needs of the project fieldwork. Several examples were provided in this book. More and more field instruments are becoming digital (e.g. data dictionaries), replacing hardcopy data collection forms. It is likely that this trend will continue. Still, many situations benefit from and require the use of hardcopy documents. Even though nearly all GPS receivers have an internal memory, having hardcopy forms that provide spaces to record information about target features is a good insurance policy. Remember to consider the cost and time involved in creating and reproducing field instruments.

Logistics

Coordination of any fieldwork always entails some set of logistical and administrative challenges. These challenges vary with such things as the complexity of the research questions, the number of people involved in fieldwork, the distance to and accessibility of the study area, the number of target features and the distance between features, and the difficulty and safety of the environment. Practically speaking, getting people and equipment to and from the study area, lodging and meals, scheduling of data collection, meetings and data download, GPS maintenance, and health and safety precautions are just some of the logistical considerations to be made in fieldwork planning and preparations.

Training

Regardless of the number of people in a project, it is important that the GPS receivers and other equipment and field instruments be used properly and data collection protocols are clearly understood and followed. Therefore, training is a very important, necessary step. Before fieldwork begins, every member of the data collection team should understand how GPS works, how to operate the specific receiver being used, and the data collection protocols. Those personnel with specialized duties, such as the field technician, will need to be trained for additional tasks, such as data downloading and data verification. Training may also consist of familiarization with the study area, if necessary, including the physical and cultural geography, local customs and taboos.

Data Collection

If all of the above steps have been done correctly and provided all equipment is in proper working order, the data collection process should run smoothly. Practical guidelines for setting up and operating GPS receivers for a number of different data collection scenarios were provided. Choose and adapt the one or more that are most applicable to the research. Keep in mind that the initial data collection may have some bumps and unforeseen problems. This should be expected. It is necessary during this phase to make sure that the data collection process does not cease when problems arise. Equipment failures should be resolved by having alternates on hand while the field technician can address other minor technical problems.

If possible, data should be regularly downloaded, organized, and backed up. Data should always be regularly verified to make certain that collection protocols are being followed, and field teams should be periodically debriefed to identify problems with equipment, protocols, morale, etc. A thorough data and equipment inventory and preparation of everything for transport out of the field concludes the fieldwork.

Data Post-Processing

After fieldwork and in order to prepare data for use in a GIS, it is first necessary to clean and review the GPS data and any attribute data for accuracy and completeness. It is important to be certain that all the coordinates that should have been collected were indeed collected, that they have been properly downloaded in the appropriate coordinate system and datum, and are properly formatted. Manual data entry should be scrutinized via cross-checks of the entered data. If post-processing differential correction was selected as an error correction technique, then the corresponding base station files will need to be acquired and used to correct the rover files. Guidelines for processing raw, averaged, and differentially correctable files should be followed in addition to the steps for turning GPS data into GIS data.

Data Integration and Analysis using GIS

Here is the payoff for all of the planning and hard work. The resulting coordinates and any accompanying attributes can be brought into a GIS environment for mapping and spatial analysis in order to answer the research questions and hopefully provide more insights and discoveries regarding the phenomenon under study. The specific analyses will of course depend on the research goals and objectives of the project, but could include distance matrices, point in polygon analysis, or integration with remotely sensed data or other spatial and nonspatial data.

Data Quality and Confidentiality

Efforts to preserve data quality should permeate all stages of project development and implementation, particularly the data collection and data entry phases, when data quality is most at risk. Data is an investment in time and resources, and a number of ways for insuring data quality were discussed in this book. They include field testing of protocols and field instruments, consistent methods, establishing data quality controls and standards, proper data management with regular download and backup and data documentation using established metadata standards. Data verification should occur at multiple times during and following fieldwork. Most importantly, researchers need to recognize the significance of data confidentiality and the potential consequences for breaking that confidence, particularly when measuring, reporting, and sharing the locations of and information about individuals and sensitive or valuable natural and cultural resources.

The Future of GPS

Undoubtedly, the popularity of GPS will continue to grow and expand into other areas of life in the coming years. Receivers will get smaller and be included in a variety of consumer products. As this book is written, GPS receivers can be found in cellular phones, digital cameras, and watches. There are also small GPS receivers available that plug into PDAs, handheld PCs, and laptop computers. Of even greater significance will be the development of GPS signals and receivers in the next couple decades that do not require a clear view of the sky to calculate position, making GPS use in difficult environments, under harsh conditions and even indoors a very real possibility.

There is great potential for these advances in GPS technology to benefit social science researchers. As receivers become smaller, cheaper, more accurate and adaptable, and are directly integrated with computers and other electronic devices, it becomes even easier to incorporate GPS into a greater number and variety of research projects. But regardless of the physical shape, size and capabilities of the receiver, developing a good foundation of knowledge, thorough project planning, and informed implementation are crucial necessities.

Summary

This book was written because the authors feel that many social science researchers can benefit from using GPS. There are many advantages to the technology; it is simple, reliable, accurate and can provide a valuable piece of information—location. However, as with any tool, if it is applied incorrectly, it is likely to produce unsatisfactory results. The information presented in this field guide is designed to give readers the knowledge necessary to properly and effectively use the tool in a variety of research settings. To return to the analogy presented at the beginning of this chapter, there are many things to be considered before getting behind the wheel of a car. However, once the strengths and limitations are explained and understood, it becomes a matter of using common sense, defining a destination, and navigating a path on the journey. It is at this stage that a book about driving a car would advise its reader, "Drive on!"

References

Ali, M., Emch, M., Tofail, F., and Baqui, A. H. 2001: Implications of health care provision on acute lower respiratory infection mortality in Bangladeshi children. *Social Science and Medicine*, 52 (2), 267–77.

Axelrad, P., and Brown, R. G. 1996: GPS Navigation Algorithms. In B. W. Parkinson and J. J. Spilker (eds), *The Global Positioning System: Theory and Application*, Vol. 1, Washington, D.C.: American Institute of Aeronautics and Astronautics, 409–33.

Clarke, K. 1995: *Analytical and Computer Cartography*. 2nd edn, Upper Saddle River, NJ: Prentice Hall.

Clarke, K. 1999: *Getting Started with Geographic Information Systems*. Upper Saddle River, NJ: Prentice Hall.

Congalton, R.G. and Green, K., 1999: *Assessing the Accuracy of Remotely Sensed Data: Principles and Practices*. Boca Raton, FL: Lewis Publishers.

Corcoran, W., 1995: Q & A: Industry Experts Answer Your GPS Questions, *Earth Observation Magazine*, May 1995. Available at: http://www.eomonline.com/Common/Archives/May95/gps.htm.

Donohue, M. J., Boland, R. C., Sramek, C. M., and Antonelis, G. A. 2001: Derelict fishing gear in the Northwestern Hawaiian Islands: Diving surveys and debris removal in 1999 confirm threat to coral reef ecosystems. *Marine Pollution Bulletin*, 21 (12), 1301–12.

Faust, K., Entwisle, B., Rindfuss, R. R., Walsh, S. J., and Sawangdee, Y. 1999: Spatial arrangement of social and economic networks among villages in Nang Rong District. *Social Networks*, 21 (4), 311–37.

Federal Geographic Data Committee, 1998: *Content Standard for Digital Geospatial Metadata*, FGDC-STD-001-1998, FGDC–United States Geological Survey, Reston, Virginia. Available at: http://www.fgdc.gov/metadata/contstan.html.

Fox, J., Yonzon, P., and Podger, N. 1996: Mapping conflicts between biodiversity and human needs in Langtang National Park, Nepal. *Conservation Biology*, 10 (2), 562–9.

Frizzelle, B. G., and McGregor, S. J. 1999: Integrating Geographic Information

Science (GISc) techniques in the data collection phase of population–environment research. In *Proceedings of the Applied Geography Conference*, Charlotte, NC: University of North Carolina at Charlotte, 199–207.

Ganskopp, D. 2001: Manipulating cattle distribution with salt and water in large arid-land pastures: a GPS/GIS assessment. *Applied Animal Behavior Science*, 73, 251–62.

Garmin Corporation, 1998: *GPS12XL Personal Navigator Owner's Manual & Reference*, Part Number 190-00134-10 Rev. A. Olathe, KS: Garmin Corp.

Garmin Corporation. 2000: *GPS Guide for Beginners*. Available at: http://www.garmin.com/manuals/gps4beg.pdf.

Gilbert, C. 1995a: GPS consumer series: Averaging GPS data without applying differential correction. *Earth Observation Magazine*, February 1995. Available at: http://www.eomonline.com/Common/Archives/Feb95/gilbert.htm.

Gilbert, C. 1995b: GPS consumer series: How to do differential GPS without known coordinates for your base station. *Earth Observation Magazine*, April 1995. Available at: http://www.eomonline.com/Common/Archives/April95/gilbert.htm

Guptill, S. C., and Morrison, J. L., 1995: *Elements of Spatial Data Quality*, Oxford: Elsevier.

Halme, M., and Tomppo, E. 2001: Improving the accuracy of multisource forest inventory estimates by reducing plot location error – a multicriteria approach. *Remote Sensing of Environment*, 78, 321–7.

Heywood, I., Cornelius, S., and Carver, S. 1998: *An Introduction to Geographical Information Systems*. Upper Saddle River, NJ: Prentice Hall.

Iliffe, J. C. 2000: *Datums and Map Projections for Remote Sensing, GIS, and Surveying*. New York: CRC Press.

Interagency GPS Executive Board. 1997: *IGEB Charter*. Available at: http://www.igeb.gov/charter.shtml.

Kammerbauer, J., and Ardon, C. 1999: Land use dynamics and landscape change pattern in a typical watershed in the hillside region of central Honduras. *Agriculture, Ecosystems and Environment*, 75 (1–2), 93–100.

Kevany, M. J. 1994: Use of GPS in GIS data collection. *Computers, Environment and Urban Systems*, 18(4), 257–63.

Lang, L., and Speed, V. 1990: A new tool for GIS. *Computer Graphics World*, 13 (10), 40–8.

Lee, J., and Mannering, F. 2002: Impact of roadside features on the frequency and severity of run-off-roadway accidents: an empirical analysis. *Accident Analysis and Prevention*, 34, 149–61.

Longley, P. A., Goodchild, M. F., Maguire, D. J., and Rhind, D. W., 2001: *Geographic Information Systems and Science*. Chichester, UK: John Wiley & Sons, Ltd.

Lumsden, L. F., Bennett, A. F., and Silins, J. E., 2002: Selection of roost sites by the lesser long-eared bat (*Nyctophilus geoffroyi*) and Gould's wattled bat (*Chalinolobus gouldii*) in a fragmented landscape in south-eastern Australia. *Journal of Zoology*, 257 (2), 207–18.

McCracken, S. D., Brondizio, E. S., Nelson, D., Moran, E. F., Siquiera, A. D., and

Rodriguez-Pedraza, C. 1999: Remote sensing and GIS at farm property level: demography and deforestation in the Brazilian Amazon. *Photogrammetric Engineering & Remote Sensing*, 65 (11), 1311–20.

Moran, E. F., Brondizio, E. S., Tucker, J. M., da Silva-Forsberg, M. C., McCracken, S. D., and Falesi., I. 2000: Effects of soil fertility and land-use on forest succession in Amazônia. *Forest Ecology and Management*, 139 (1–3), 93–108.

NIMA 1997: *NIMA Technical Report TR8350.2, Department of Defense World Geodetic System 1984, Its Definition and Relationships with Local Geodetic Systems*, Third Edition, 4 July 1997, Washington, D.C.: National Imagery and Mapping Agency, US Department of Defense.

OSGB 2001: *A Guide to Coordinate Systems in Great Britain*, Southampton, U.K.: Ordnance Survey of Great Britain. Available at: http://www.gps.gov.uk/guidecontents.asp.

Parkinson, B. W. 1996a: Introduction and Heritage of NAVSTAR. In B.W. Parkinson and J.J. Spilker (eds), *The Global Positioning System: Theory and Application*, Vol. 1, Washington, D.C.: American Institute of Aeronautics and Astronautics, 3–28.

Parkinson, B. W. 1996b: GPS Error Analysis. In B.W. Parkinson and J.J. Spilker (eds), *The Global Positioning System: Theory and Application*, Vol. 1, Washington, D.C.: American Institute of Aeronautics and Astronautics, 469–83.

Pendleton, G. 2002: GPS/GIS Integration: A Buyer's Guide for Savvy GPS Consumers. *GEOWorld*, 15 (1), 24–5. Available at: http://www.geoplace.com/gw/2002/0201/0201gnt.asp

Perry, B., and Gesler, W. 2000: Physical access to primary health care in Andean Bolivia. *Social Science and Medicine*, 50 (9), 1177–88.

Priskin, J. 2001: Assessment of natural resources for nature-based tourism: the case of the Central Coast Region of Western Australia. *Tourism Management*, 22, 637–48.

Robinson, A. H., Morrison, J. L., Muehrcke, P. C., Kimerling, A. J., and Guptill, S. C. 1995: *Elements of Cartography*. 6th edn, New York: John Wiley & Sons, Inc.

Rybaczuk, K. Y. 2001: GIS as an aid to environmental management and community participation in the Negril Watershed, Jamaica. *Computers, Environment and Urban Systems*, 25 (2), 141–65.

Seagle, D. E., and Bagwell, L. V. 2001: Mapping Blackfeet Indian Reservation irrigation systems with GPS and GIS. *Photogrammetric Engineering & Remote Sensing*, 67 (2), 171–8.

Sexton, W. T., Dull, C. W., and Szaro, R. C. 1998: Implementing ecosystem management: a framework for remotely sensed information at multiple scales. *Landscape and Urban Planning*, 40 (1–3), 173–84.

Slater, J. A. and Malys, S. 1997: WGS84–past, present, and future. In F. K. Brunner (ed.), *Advances in Positioning and Reference Frames, Proceedings of the IAG Scientific Assembly, September 3–9, 1997*, New York: Springer-Verlag.

Stem, J. E. 1989: *State Plane Coordinate System of 1983*, Washington D.C.: U.S. Department of Commerce.

Tobler, W. 1970: A computer movie. *Economic Geographer*, 46, 234–40.

Trimble Navigation Limited. 1996: *Mapping Systems: General Reference.* Sunnyvale, CA: Trimble Navigation Limited.

United States Department of Commerce. 1998: *Global Positioning System, Market Projections and Trends in the Newest Global Information Utility,* The International Trade Administration, Office of Telecommunications, U.S. Department of Commerce.

United States Department of Defense. 1991: *GPS NAVSTAR User's Overview,* Prepared for the Program Director, NAVSTAR Global Positioning System, Joint Program Office, ARINC Research Corporation, Los Angeles, CA. Document No. YEE-82-009D.

United States Geological Survey. 1947: United States National Map Accuracy Standards, In *National Mapping Program Standards,* 3rd Version, 17 June 1947. Available at: http://rockyweb.cr.usgs.gov/nmpstds/nmas647.html.

United States Geological Survey, 1999: *Map Accuracy Standards, Factsheet FS-171-99,* November, U.S. Department of the Interior–USGS, Reston, Virginia. Available at: http://mac.usgs.gov/mac/isb/pubs/factsheets/fs17199.html.

Van Sickle, J. 2001: *GPS for Land Surveyors.* Second edition. Chelsea, MI: Ann Arbor Press.

Wilkinson, D., and Tanser, F. 1999: GIS/GPS to document increased access to community-based treatment of tuberculosis in Africa. *The Lancet,* 354 (9176), 394–5.

Appendix A
GPS Manufacturers

Benefon
Benefon Oyj (Head Office/Factory), P.O. Box 84, Meriniitynkatu 11, FIN-24101 Salo, Finland
www.benefon.com

Brunton
Brunton, 620 East Monroe Avenue, Riverton, WY 82501, USA
www.brunton.com

Corvallis
Corvallis Microtechnology, Inc., 413 SW Jefferson Ave.. Corvallis, OR 97333, USA
www.cmtinc.com

Garmin International
1200 East 151st Street, Olathe, KS 66062, USA
www.garmin.com

Leica
2801 Buford Highway North East, Suite 300, Atlanta, GA 30329, USA
www.leica-geosystems.com

Lowrance
12000 East Skelly Drive, Tulsa, OK 74128, USA
www.lowrance.com

Magellan
471 El Camino Real, Santa Clara, CA 95050—4300, USA
www.magellangps.com

Navman
NAVMAN NZ Limited
13–17 Kawana Street, Northcote, PO Box 68 155 Newton, Auckland, New Zealand
www.navman.com

Standard Horizon
Standard Horizon, Marine Division of Vertex Standard, 10900 Walker Street, Cypress, CA 90630, USA
www.standardhorizon.com

Thales
Thales Navigation, 471 El Camino Real, Santa Clara, CA 95050–4300 USA
www.thalesnavigation.com

Trimble
645 North Mary Avenue, Sunnyvale, CA 94088, USA
www.trimble.com

Appendix B
Sample Field Instrument

This appendix contains a blank sample field instrument. This field instrument has been designed as a multipurpose tool for use in most types of field data collection campaigns. Its purpose is to serve as a record for the GPS data collected, whether it is used to attribute data after the fieldwork is complete or it is stored as a backup file in case of loss of data. It is important to note, however, that this is simply one way of designing such an instrument.

To briefly explain the different aspects of the field instrument, begin at the very top. The first section contains a field to enter the *Site ID*, which is some predetermined ID that designates the GPS data being collected. The field next to it is for the *Date*.

Directly to the right of the *Date* field are several fields where the user can enter information about the coordinate system and horizontal datum being used to collect the GPS data, as well as the distance units and any other information pertinent to the data collection.

Below *Site ID* and *Date* is a section in which the types of features whose locations are being captured with the GPS are indicated. The user should first check the box to indicate if the file being collected represents a *Point*, *Line*, or *Polygon*. Secondly, the user checks a box to indicate if the feature is a *Building* (such as a house or facility), some *Physical or Land Use/Land Cover* (LULC) feature, a *Road*, or *Other*. If *Other* is checked, a brief description can be written in the remaining area.

The next two sections, below and to the right of the *Type of Feature* section, are where the actual file information is entered. If the user is only collecting a waypoint, such as in cases where a recreational receiver is being used, only the *Waypoint Name*, *Waypoint X*, and *Waypoint Y* need be entered. However, if the user is using a mapping grade receiver to collect rover files, then the *Rover File Name*, *Rover # of Points*, and *Rover File Size* should be entered. The *Rover # of Points* is simply the number of points that were collected during the time that the rover file was open. The *Rover File Size* is the size, often in kilobytes (kb), of the file itself. Both of these bits of information may seem meaningless, but they are extremely useful when verifying that a rover file name on a field sheet matches a file name on a computer. The number of points and the file size can help identify mismatches and errors in transcription.

The last several sections are for additional contextual information that will help the end users effectively utilize the GPS data in their work. First is an area for a *General Description* of the feature. Next to that is a field for describing the *Location of GPS Point*. For example, if a point was collected at the front door to represent a house, then in this field the user can write, "collected at the front door." Below these sections are areas for describing GPS files taken along roads. Information on *Surface Type* and *Travel Time/ Speed* can help end users attribute GIS road files for use in analyses such as network analysis (see Chapter 13). The next section is an area in which the user can draw a sketch of the feature. In the example above, the user could draw a rectangle representing the house (view from above), indicate the location of the door, draw an "X" to represent the GPS point, and finally indicate which direction is north, if deemed necessary. This *Sketch* area is broken up into four quadrants for use in GPS collection of physical or LULC data. The user can sketch out the features of the landscape in four quadrants around the GPS point, or can sketch the contents of the area if a GPS polygon is collected. The final section is provided for any other comments or information that the user deems necessary to record on the sheet.

GPS Field Collection Log Sheet			
Site ID:	Date:	Coordinate System:	Horizontal Datum:

Type of Feature:
☐ Point   ☐ Line   ☐ Polygon
☐ Building   ☐ Physical/LULC   ☐ Road
☐ Other:

Units:

Other Coord. Sys. Information:

Waypoint Name:

Rover File Name:

Waypoint X:

Rover # of Points:	Rover File Size:	Waypoint Y:

General Description:

Location of GPS Point:

Surface Type:

Travel Time/Speed

Sketch:

Other:

Appendix C
UTM Zones

Zone	Central Meridian	Extent	Zone	Central Meridian	Extent
1	177° W	180° W – 174° W	28	15° W	18° W – 12° W
2	171° W	174° W – 174° W	29	9° W	12° W – 6° W
3	165° W	169° W – 162° W	30	3° W	6° W – 0°
4	159° W	162° W – 156° W	31	3° E	0° – 6° E
5	153° W	156° W – 150° W	32	9° E	6° E – 12° E
6	147° W	150° W – 144° W	33	15° E	12° E – 18° E
7	141° W	144° W – 138° W	34	21° E	18° E – 24° E
8	135° W	138° W – 132° W	35	27° E	24° E – 30° E
9	129° W	132° W – 126° W	36	33° E	30° E – 36° E
10	123° W	126° W – 120° W	37	39° E	36° E – 42° E
11	117° W	120° W – 114° W	38	45° E	42° E – 48° E
12	111° W	114° W – 108° W	39	51° E	48° E – 54° E
13	105° W	108° W – 102° W	40	57° E	54° E – 60° E
14	99° W	102° W – 96° W	41	63° E	60° E – 66° E
15	93° W	96° W – 90° W	42	69° E	66° E – 72° E
16	87° W	90° W – 84° W	43	75° E	72° E – 78° E
17	81° W	84° W – 78° W	44	81° E	78° E – 84° E
18	75° W	78° W – 72° W	45	87° E	84° E – 90° E
19	69° W	72° W – 66° W	46	93° E	90° E – 96° E
20	63° W	66° W – 60° W	47	99° E	96° E – 102° E
21	57° W	60° W – 54° W	48	105° E	102° E – 108° E
22	51° W	54° W – 48° W	49	111° E	108° E – 114° E
22	45° W	48° W – 42° W	50	117° E	114° E – 120° E
24	39° W	42° W – 36° W	51	123° E	120° E – 126° E
25	33° W	36° W – 30° W	52	129° E	126° E – 132° E
26	27° W	30° W – 24° W	53	135° E	132° E – 138° E
27	21° W	24° W – 18° W	54	141° E	138° E – 144° E

University of Glamorgan
Learning Resources Centre -
Treforest
Self Issue Receipt

Customer name: MR JEROME
LOUIS THOMAS
Customer ID: ******0972302

Title: Global Positioning System : a
field guide for the social sciences
ID: 7312484571
Due: 15/05/2008 23:59

Total items: 1
17/04/2008 23:07

Thank you for using the Self-Service
System
Diolch yn fawr

Zone	Central Meridian	Extent
55	147° E	144° E – 150° E
56	153° E	150° E – 156° E
57	159° E	156° E – 162° E
58	165° E	162° E – 168° E
59	171° E	168° E – 174° E
60	177° E	174° E – 180° E

Index